Research on the Crack Control of Concrete

Research on the Crack Control of Concrete

Guest Editors

Lepeng Huang
Zuowei Liu
Lin Chen

MDPI

Basel • Beijing • Wuhan • Barcelona • Belgrade • Novi Sad • Cluj • Manchester

Guest Editors
Lepeng Huang
School of Civil Engineering
Chongqing University
Chongqing
China

Zuowei Liu
School of Civil Engineering
Yantai University
Yantai
China

Lin Chen
Department of Public and International Affairs (PIA)
City University of Hong Kong
Hong Kong
China

Editorial Office
MDPI AG
Grosspeteranlage 5
4052 Basel, Switzerland

This is a reprint of the Special Issue, published open access by the journal *Buildings* (ISSN 2075-5309), freely accessible at: https://www.mdpi.com/journal/buildings/special_issues/MAWWM8V49W.

For citation purposes, cite each article independently as indicated on the article page online and as indicated below:

Lastname, A.A.; Lastname, B.B. Article Title. *Journal Name* **Year**, *Volume Number*, Page Range.

ISBN 978-3-7258-6556-7 (Hbk)
ISBN 978-3-7258-6557-4 (PDF)
https://doi.org/10.3390/books978-3-7258-6557-4

Contents

About the Editors

Lepeng Huang

Lepeng Huang is an Associate Researcher and Deputy Director of the Institute of Construction Engineering at the School of Civil Engineering, Chongqing University. He also serves as the Executive Deputy Director of the Modern Construction Technology Research Center. He has been recognized as a New Chongqing Youth Innovation Talent, an Outstanding Young Professional by the Chongqing Civil Engineering Society, and an expert member of several national and industry committees, including the Expert Committee of the China Construction Industry Association, the Standardization Committee of the China Construction Enterprise Management Association, and the Expert Panel of the Chongqing Construction Industry Association. He has led three national research projects and contributed to five additional national and provincial-level projects. He has published 42 SCI-indexed papers, including eight highly cited articles, and holds more than 60 granted invention patents and over 20 utility model patents. He has served as a lead or contributing drafter of three national standards, ten local standards, and two association standards. His research achievements have been recognized with more than ten awards, including the First Prize of the Chongqing Science and Technology Progress Award and the First Prize of the China State Construction Engineering Corporation Science and Technology Award. He has participated extensively in scientific and technological work for major engineering projects. He has led or participated in five teaching reform projects, published four textbooks, and contributed to one national-level virtual simulation experimental teaching first-class course and one municipal-level online first-class course. He has supervised more than ten student innovation and entrepreneurship projects and has received several teaching honors.

Zuowei Liu

Zuowei Liu is a Lecturer at the School of Civil Engineering, Yantai University. His research program is centered on the development of sustainable and high-performance construction materials, with a specific focus on green concrete technology. Dr. Liu's work encompasses several key areas, including the enhancement of mechanical properties and durability in eco-friendly concrete, the application of advanced biomineralization techniques in building materials, and the comprehensive environmental assessment of construction products through life-cycle analysis (LCA). He has established a strong record of scholarly output, having authored and published multiple high-impact papers in prestigious, high-ranking SCI journals. His notable publications include significant research on the biological modification of recycled aggregate concrete and innovative methods for improving its resistance to freeze-thaw cycles, contributing valuable knowledge to the field. In addition to his fundamental research, Dr. Liu actively engages in applied research and development with strong industry relevance. He has successfully secured and led several industry-sponsored applied research (horizontal) projects. These projects are directed towards the practical development and optimization of novel building materials and technologies. Dr. Liu also contributes to the broader academic community by serving as a reviewer for several top international journals such as *Construction and Building Materials* and the *Journal of Cleaner Production*. Through his combined efforts in pioneering academic research, impactful industry collaboration, and dedicated professional service, Dr. Liu is committed to providing innovative and sustainable engineering solutions that address contemporary challenges in civil engineering and construction materials science.

Lin Chen

Lin Chen is a Research Associate at City University of Hong Kong. His research focuses on green construction, life-cycle assessment, and low-carbon building materials. He received his Ph.D. from Chongqing University and conducted joint doctoral research at the University of Cambridge. Dr. Chen is a member of several international professional organizations, including the Institution of Civil Engineers (ICE), the American Society of Civil Engineers (ASCE), and the Chartered Institute of Building (CIOB). He has been recognized in the global top 2% of scientists in both 2024 and 2025. He has published 28 SCI-indexed papers, including 10 papers listed among the top 1% highly cited and 0.1% hot papers in ESI. As of December 2025, his publications have received approximately 6,200 citations on Google Scholar, with an H-index of 20. He holds 15 patents and software copyrights. He serves as an Assistant Editor for the journal *Engineering* and acts as a reviewer for multiple SCI journals. He has participated in more than 20 national, provincial, and university-level research projects, including those funded by the National Natural Science Foundation of China. He has received numerous honors, such as the National Scholarship and the Baosteel Scholarship.

Preface

The field of concrete materials continues to evolve as new experimental techniques, analytical formulations, and computational tools deepen our understanding of cracking behavior, durability, and structural performance under increasingly complex engineering service conditions. This Reprint brings together a series of contributions that reflect ongoing research efforts dedicated to advancing both the fundamental principles and the practical applications of concrete and cementitious materials.

The scope of this Reprint spans early-age shrinkage, moisture diffusion behavior, mixed-mode fracture, probabilistic modeling, structural cracking assessment, and the multi-physics coupling processes that govern long-term performance. These studies collectively highlight the importance of integrating theoretical development with experimental verification and numerical simulation in order to capture the complex interactions that control deformation, damage, and failure in concrete structures.

The aim of this Reprint is to provide readers with a coherent overview of the current progress in the discipline and to present representative research that illustrates the latest methodological advances. It also seeks to promote the exchange of ideas across different subfields, including material science, structural engineering, durability assessment, and computational mechanics. The Reprint is intended for researchers, engineers, and graduate students who are engaged in the study of concrete materials and structural behavior, as well as those interested in the development of predictive models and design-oriented evaluation methods.

We hope that the collection of articles assembled in this Reprint will serve as a useful reference and stimulate further exploration into the complex mechanisms that influence the performance of concrete in modern engineering practice.

Lepeng Huang, Zuowei Liu, and Lin Chen
Guest Editors

buildings

MDPI

Article

Early Shrinkage Modeling of Complex Internally Confined Concrete Based on Capillary Tension Theory

Fengbin Zhou [1], Hao Jiang [1], Lepeng Huang [1,*], Ying Hu [1], Zhuolin Xie [1], Zhikai Zeng [2], Maoyi Liu [2], Bo Wang [3] and Xingyang Zhou [4]

1 School of Civil Engineering, Chongqing University, Chongqing 400045, China; 20141613158@cqu.edu.cn (F.Z.); 201716021056@alu.cqu.edu.cn (H.J.); huying@stu.cqu.edu.cn (Y.H.); zhuolin.xie@cqu.edu.cn (Z.X.)
2 Chongqing Urban Investment Infrastructure Construction Co., Ltd., Chongqing 400045, China; m18290409197@163.com (Z.Z.); liumaoyi@163.com (M.L.)
3 China Metallurgical Construction Engineering Group Co., Ltd., Chongqing 400084, China; ys102412@163.com
4 Department of Civil Engineering, Xi'an Jiaotong-Liverpool University, Suzhou 215123, China; xingyang.zhou20@student.xjtlu.edu.cn
* Correspondence: huang_lepeng@cqu.edu.cn

Abstract: This paper evaluates the shrinkage performance of concrete under complex internal constraint environments comprising steel plates, studs, and reinforcement to investigate their respective influence laws on the shrinkage performance of concrete. An early shrinkage model of concrete under complex internal constraints was established based on the theory of capillary tension, and the effects of steel plate, nails, and steel reinforcement on the shrinkage performance of concrete were theoretically analyzed. Six sets of concrete-constrained shrinkage tests and pore structure tests were then performed under different internal constraint conditions with the steel plate thickness, reinforcement diameter, and stud-related parameters (stud diameter, height, and spacing) as research variables. The test results demonstrate that the pore structure of concrete increases with the increase in the constraint coefficient, and that the increase in the pore structure will cause a decrease in the capillary pore stress, which is the driving force of concrete shrinkage. Its decrease will inevitably lead to a decrease in concrete shrinkage. By comparing the calculated values of the shrinkage model with the measured values, it is found that the average value of the prediction error is less than 15%, which reveals that the predicted values of shrinkage are in good agreement with the measured values and proves that the model can effectively predict the shrinkage of concrete that is restrained by steel plates, pins, and reinforcing bars.

Keywords: shrinkage modeling; internal constraints; capillary pore stress; steel plate–concrete composite shear wall

1. Introduction

In recent years, steel plate–concrete shear walls have been widely used in buildings in high-intensity seismic zones because of their full hysteresis curve and high energy dissipation capacity [1–3]. The structural system in which they are used is very wide and can be mainly categorized into the following four structural systems: frame-core, frame tube-core, mega-frame-core, and mega-frame-core-mega bracing [4]. The core portion in the above four systems is generally composed of steel–concrete combined shear walls. Despite its engineering applications, the tendency of the steel–concrete combined shear wall toward shrinkage cracking is being gradually realized. Researchers in various countries have also conducted extensive research on steel–concrete combined shear wall cracking problems over the past decades, but the main direction of research is the structural form of the steel–concrete combined shear wall as well as research into their force performance [5,6]. Existing steel–concrete combined shear wall cracking analysis and control is based on the reinforced concrete structure of cracking as established from actual projects. However, this cracking control method is unsatisfactory, steel–concrete combined shear wall cracks occur prominently in the early stage as the concrete cracks at that stage severely affect the waterproofing of the building, leading to corrosion and other effects that result in these early-stage steel–concrete combined shear wall cracks [7]. This further reduces the structural bearing capacity of the steel–concrete combined shear wall, and, if the cracks progress beyond a certain extent, the durability of the structure will be fully compromised. Theoretically, the steel plate–concrete combination shear wall withstands less than the normal load in the early stage, so its early cracking is mostly caused by deformation. Non-load cracks occur in concrete at an early stage mainly due to the early shrinkage of concrete and the limitation in its shrinkage due to the corresponding internal and external constraints.

As the water in the concrete is gradually consumed by hydration and drying, capillary pore stresses will occur in the pores of the concrete [8]. The stresses cause the distance between particles within the concrete to decrease, resulting in a macroscopic decrease in the volume of the concrete, a phenomenon known as shrinkage [9]. Concrete shrinkage deformation in the early stage is subjected to the external constraints of the neighboring components. The concrete will be subjected to constrained tensile stresses internally because the internal reinforcing bars, steel plates, and other constraining components will not produce shrinkage deformation. Meanwhile, because the early tensile strength of the concrete is not high, when the constrained tensile stress exceeds the tensile strength of concrete, cracks are produced [10,11]. On the one hand, from the point of view of shrinkage deformation, and due to the high structural performance requirements for steel plate–concrete combined shear walls in many projects, high-strength concrete (C60 and above strength grade) is often used in the steel plate–concrete combined shear wall. The water–cement ratio of high-strength concrete is lower than that in weaker concretes, leading to the resulting shrinkage deformation in its early stage being more significant than that of ordinary concrete [12]. On the other hand, from the perspective of internal constraints, the steel plate–concrete combined shear wall is arranged with dense reinforcement, bolts, and steel plates, which leads to stronger internal constraints for the contraction deformation of high-strength concrete [13]. Under such strong internal constraints, high-strength concrete is subject to stronger internal constraints on tensile stresses, which is one of the reasons for the cracking problem in steel–concrete combined shear walls.

Researchers have conducted a series of studies on the confined shrinkage of concrete. For example, Doo-YeolYoo et al. [14] studied the effects of shrinkage reducers and different reinforcement ratios on the shrinkage properties of ultra-high performance fiber-reinforced concrete (UHPFRC). The results of the study revealed that the shrinkage stress of UHPFRC with a low reinforcement ratio was lower, and using more shrinkage reducers could effectively reduce the shrinkage stress of UHPFRC. Shen Dejian et al. [15] conducted a related study on the effects of curing temperature on the shrinkage properties of high-strength concrete. Their results demonstrate that the shrinkage deformation of high-strength concrete under 100% constraint increased with the increase in curing temperature, and the risk of cracking increased. Ehsan Negahban et al. [16] studied the restrained shrinkage and creep behaviors of geopolymer concrete (GPC). Their results demonstrate that GPC had 38–57% less restrained shrinkage than normal concrete. Inamullah Khan et al. [17] conducted a series of experiments to assess the influences of early-age shrinkage on cracking in reinforced-concrete (RC) members subjected to internal restraint and found that the magnitude of the restrained shrinkage depends on the reinforcement ratio. Some researchers [18] experimentally studied the effects of various admixtures, including expansion agents, internal curing agents, and internal curing agent compounding, on the early shrinkage and cracking of concrete based on early flat-plate constrained cracking. Huang L et al. [19] investigated the effects of steel plates and pins on the shrinkage properties and cracking properties of high-strength concrete and found that the thickness of steel plates and their modulus of elasticity lead to an increase in the shrinkage and deformation of high-strength concrete. Moreover, the risk of cracking increased with the modulus of elasticity, whereas the inhibition of concrete shrinkage by the steel plate decreased with the increase in the distance between the concrete and the steel plate. Further, the cracking of high-strength concrete subjected to steel plate and studs may start from the bottom and develop along the boundary of the studs.

These results of concrete-constrained shrinkage reveal the way that scholars are becoming increasingly concerned to study the shrinkage performance of concrete under various constrained states, and the factors studied have included admixtures, curing conditions, mineral admixtures, and internal and external constraints. Further, constrained tests tend to increasingly involve uniaxial constrained tests to quantitatively analyze the shrinkage performance of constrained concrete as a method of predicting its risk of cracking [20–22]. However, the current research on the internal restraint of concrete focuses on the internal restraint of steel reinforcement by considering the influence of reinforcement rate, elastic modulus of steel reinforcement, and reinforcement method on the shrinkage performance of concrete. However, in an actual steel plate–concrete combined shear wall, not only is the internal restraint of steel reinforcement present, but the complex internal restraint of a steel plate, spigot, and steel reinforcement exists. However, research on the shrinkage performance of concrete under this type of complex internal restraint is lacking. Therefore, more research on this type of complex internal constraint is needed.

In this study, a strain analysis model of concrete under complex constraints of steel plates, bolts, and steel bars was established first based on the theory of capillary tension. Subsequently, experiments were conducted to explore the shrinkage deformation ability of concrete under internal constraints (steel plates, bolts, and steel bars). The main variables of the experiment were steel plate thickness, steel bar diameter, bolt diameter, bolt spacing, and bolt length. The experiments include basic mechanical, restrained shrinkage, and pore structure tests. Using the above data, the influence of steel plates, bolts, and steel bars on the restraint capacity of concrete shrinkage was analyzed, and the correctness and applicability of the established theoretical model were verified.

2. Experimental Design

2.1. Concrete Mixing Ratio

High-strength concrete (C60) was used to make the specimens. The cementitious materials used to configure the concrete specimens contained P.O 42.5R ordinary silicate cement and Chengdu Bolei Class I fly ash. Pebble gravel with a maximum nominal size of 31.5 mm was used as coarse aggregate, and pebble mechanism medium sand was used as fine aggregate. Zhongjian polyhydroxylic acid high-efficiency water reducer ZJC-01 was used as an additive. This is a colorless transparent or light-yellow liquid with a pH of 6–7, a water reduction rate of 40%, and an apparent viscosity of 675 mPa.S. The concrete mix ratio is presented in Table 1.

Table 1. Concrete mix ratio (kg/m^3).

Water to Binder Ratio	Cement	Course Aggregate	Sand	Water	Fly Ash	Additive
0.29	482	1064	680	154	48	12.7

2.2. Specimen Design

The specimens used in the test had the dimensions of an actual steel plate–concrete shear wall and a similar arrangement of the internal restraining members, which was designed based on the JGJT 380-2015 [23], GB 50010-2010 (2015 edition) [24], JGJ 3-2010 [25], and JGJ 138-2016 [26] with GB 50011-2010 (2016 edition) [27].

To study the effect of the steel plate, spigot, and steel reinforcement on the shrinkage properties of concrete, six internally restrained concrete specimens and one plain concrete specimen were designed based on the "basic analysis mold" proposed in Section 3.1 of this article. The height of the specimens was 1000 mm, the cross-section size was

1000 × 200 mm, and the span-to-height ratio was 1.0, whereas the size of the steel plate was 1000 × 1000 mm (see Figure 1a,b), and the S1 specimen parameter design is as follows:

1. Steel strength is Q235B, and thickness is 10 mm;
2. The diameter of the bolts is 16 mm, the length is 80 mm, and it is welded on both sides of the steel plate in a square arrangement, as shown in Figure 1d;
3. Horizontal distribution reinforcement is Φ10@200, and longitudinal horizontal reinforcement is Φ10@200. These are bidirectionally arranged along the test specimen in two layers, with the same diameter of reinforcement in both directions. The studs are passed through the distribution reinforcement mesh, as shown in Figure 1c,e.

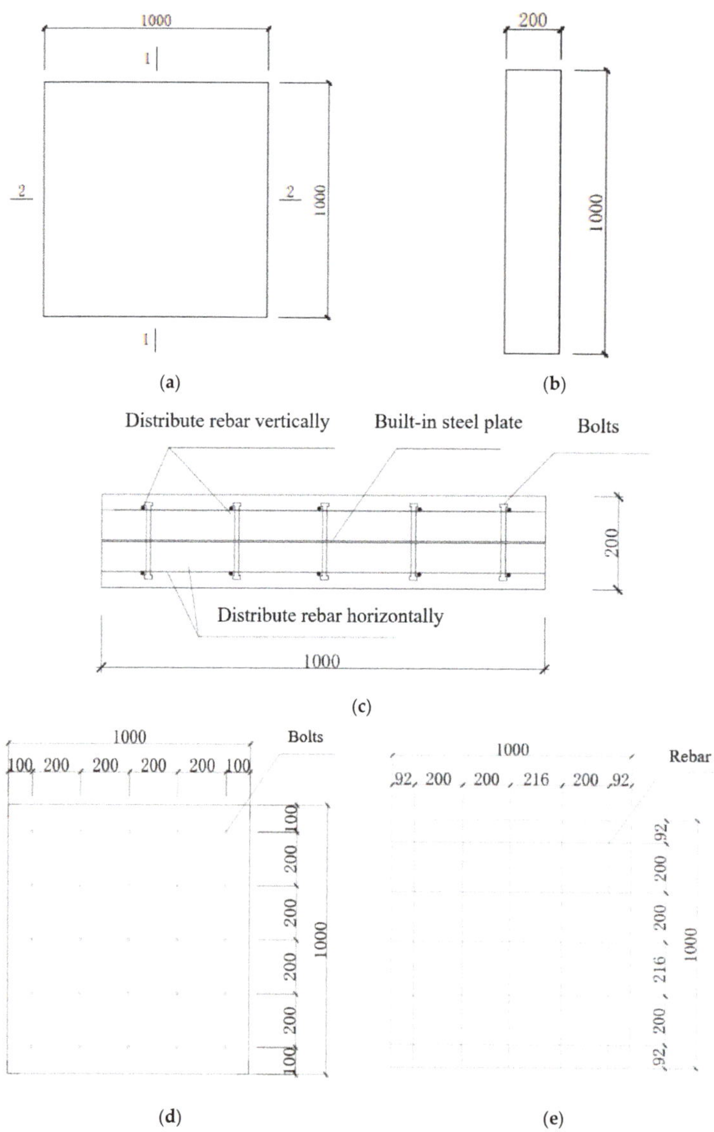

Figure 1. Size and construction of specimens: (**a**) Elevation of the test piece; (**b**) 1-1 Cross-section; (**c**) 2-2 Cross-section; (**d**) Test piece S1 bolt arrangement drawing; (**e**) Test piece S1 rebar layout drawing.

In the test, the connection between the studs and the steel plate was welded to ensure synergy between the steel plate and concrete, and the studs passed through the reinforcement network to strengthen the effective bond between the concrete and the steel plate. The thickness of the steel plate, diameter of the reinforcement bars, diameter of the studs, length of the studs, and the stud spacing were respectively changed to form six restrained concrete specimens. Finally, one plain concrete specimen was designed for the sealed curing condition. The specimen design details are shown in Table 2.

Table 2. Specimen design.

No.	Section Size (mm)	Plate Thickness (mm)	Distribution Rebar Straight Diameter (mm)	Bolt Diameter (mm)	Bolt Height (mm)	Bolt Spacing (mm)	Curing Condition
P	1000 × 200	0	0	0	0	0	Sealed
S1	1000 × 200	10	10	16	80	200	Sealed
S2	1000 × 200	12	10	16	80	200	Sealed
S3	1000 × 200	10	10	16	80	100	Sealed
S4	1000 × 200	10	10	19	80	200	Sealed
S5	1000 × 200	10	10	16	70	200	Sealed
S6	1000 × 200	10	12	16	80	200	Sealed

2.3. *Performance Testing*

2.3.1. Mechanical Properties Testing

The cubic compressive strength, split tensile strength, and static compressive elastic modulus of concrete cured for 3 d, 7 d, 14 d, and 28 d were tested. The test methods were performed according to GBT 50081-2019 [28]. The test results are listed in Table 3.

Table 3. Specimen information.

Basic Properties	Age (Days)			
	3	7	14	28
Cube compressive strength (MPa)	42.8	48.2	56.4	61.0
Splitting tensile strength (MPa)	3.8	4.1	4.5	5.2
Elastic modulus (Gpa)	30.0	32.3	35.0	35.5

2.3.2. Concrete Restraint Shrinkage Test

To measure the shrinkage deformation of concrete in the longitudinal direction, three measurement points were arranged 100 mm, 500 mm, and 900 mm from the bottom of the specimen, termed measurement points 1, 2, and 3, respectively, and the arrangement of the shrinkage measurement points is shown in Figure 2. The data were collected using the automated collection equipment, which consists of a displacement transducer (micrometer), main (sub) hub, laptop computer, and micrometer processing system, as shown in Figure 3.

When measuring the shrinkage displacement of concrete, the measuring point bolts should be pre-embedded at the corresponding positions. After the initial set of concrete is removed, the bolts should be screwed back to their original positions. Then, the digital electronic dial gauge should be installed at the corresponding measuring point position in the longitudinal direction. The dial gauge head should be in full contact with the bolt, and the position should be adjusted so that the instrument is horizontally and vertically aligned with the measuring point. Then, the dial gauge should be connected to the data line, the data cable to the diversity cable, the diversity cable to a main hub, and the main hub to the laptop using a USB adapter, with recording being conducted automatically.

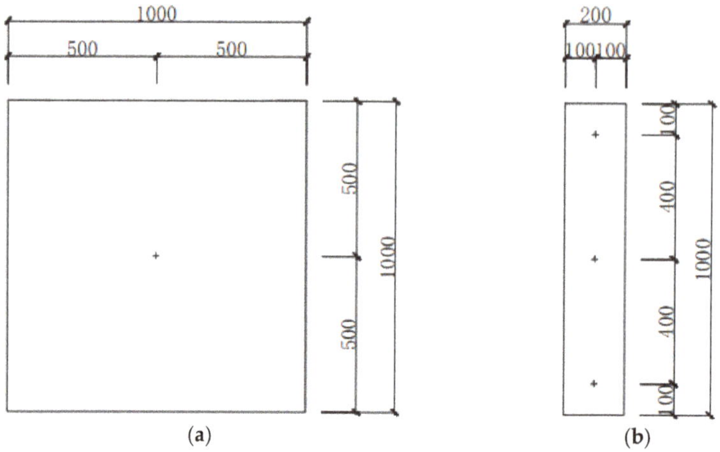

Figure 2. Arrangement of measuring points for shrinkage of restrained concrete: (**a**) elevation plan; (**b**) side view.

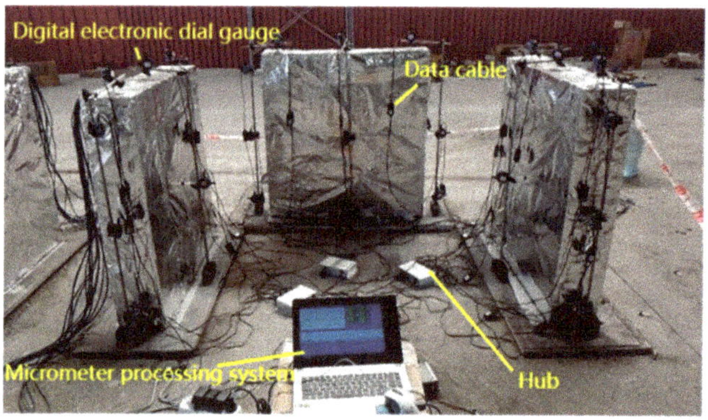

Figure 3. Automatic acquisition equipment.

2.3.3. Concrete Pore Structure Test

The specimens used for the concrete pore structure tests had the same specifications as those used for the shrinkage tests. At each test age (3, 14, and 28 d), the specimens were drilled, some samples of concrete materials were collected, and the concrete was crushed with a sieve to select concrete particles with sizes between 2.5 mm and 5 mm, ensuring that the volume of each particle was less than 1 cm^3. Subsequently, the hydration of the concrete was prevented using acetone, and the specimens were dried in a vacuum dryer. Finally, the internal pore structure of the concrete was tested using an AUTOIV9510 fully automated mercury porosimeter.

3. Shrinkage Modeling of Concrete with Complex Internal Constraints Based on Capillary Tension Theory

3.1. Shrinkage Modeling of Concrete with Complex Internal Constraints

Capillary tension theory has been shown to better explain and predict the shrinkage of plain concrete. Therefore, in this section, a model of concrete shrinkage subjected to the joint constraints of a steel plate, studs, and reinforcement is developed based on capillary tension theory. When concrete shrinks under capillary stress, the steel plate and studs themselves do not deform. Therefore, restraining shear stress is generated at the interface between the steel plate, studs, and concrete. This shear stress leads to compressive stresses in the steel plate, studs, and reinforcement, whereas restraining tensile stresses will be formed inside the concrete in the direction opposite to the direction of contraction of the concrete, as shown in Figure 4.

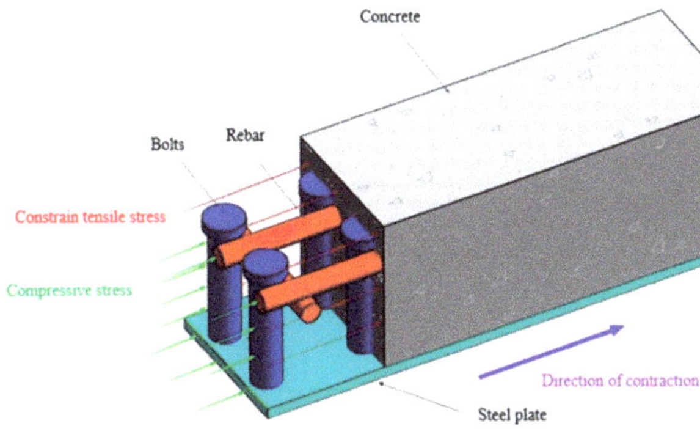

Figure 4. Basic analysis mold.

According to the above basic principles, a concrete constrained shrinkage model with a total length of L, n bolts with a diameter of d, and N reinforcement bars (divided into transverse and longitudinal reinforcement bars) with a diameter of D, is established, which is constrained by the steel plate, bolts, and reinforcement bars. A concrete unit in the constrained model is taken as the object of analysis, as shown in Figure 5.

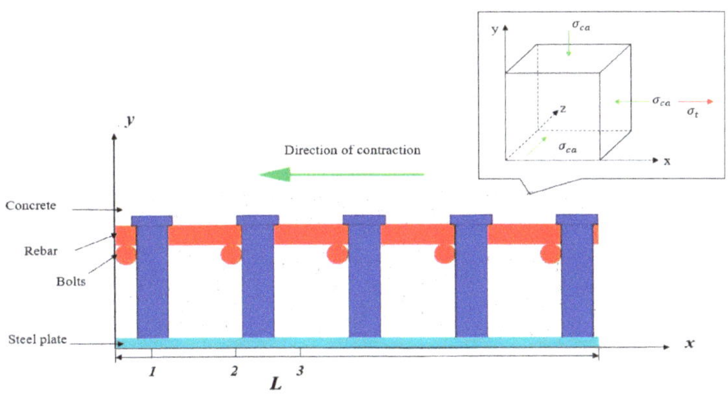

Figure 5. Concrete element analysis model.

According to the theory of capillary tension, when the water in concrete is consumed, the capillary stress generated on the walls of the concrete capillaries becomes the main driving force for the shrinkage of concrete, and the capillary stress (σ_{ca}) that causes the shrinkage of concrete can be expressed as follows:

$$\sigma_{ca} = \Delta P = \frac{2\gamma \cos\theta}{r} \tag{1}$$

where γ is the surface tension of the capillary wall, which is equal to 7.28×10^{-2} N/m at 20 °C; θ denotes the contact angle at the liquid–solid interface (0 for concrete); and r refers to the most accretive pore size of the pores.

Due to discontinuities in the arrangement of studs and reinforcement, three typical cross-sections in concrete are constrained by steel plates, studs, and reinforcement, namely, cross-section 1 with steel plates, concrete, studs, and longitudinal reinforcement;

cross-section 2 with steel plates, concrete, transverse reinforcement, and longitudinal reinforcement; and cross-section 3 with steel plates, concrete and longitudinal reinforcement, as shown in Figure 6. The total length of the model with type 1 cross-section is nd, the total length of the model with type 2 cross-section is ND, and the total length of the model with type 3 cross-section is $L - nd - ND$. To simplify the analysis, the diameters of the heads of the studs were replaced by the diameters of the stud rods in this model, and the projected area of the studs was used to represent the cross-sectional area of the studs in cross-section 1.

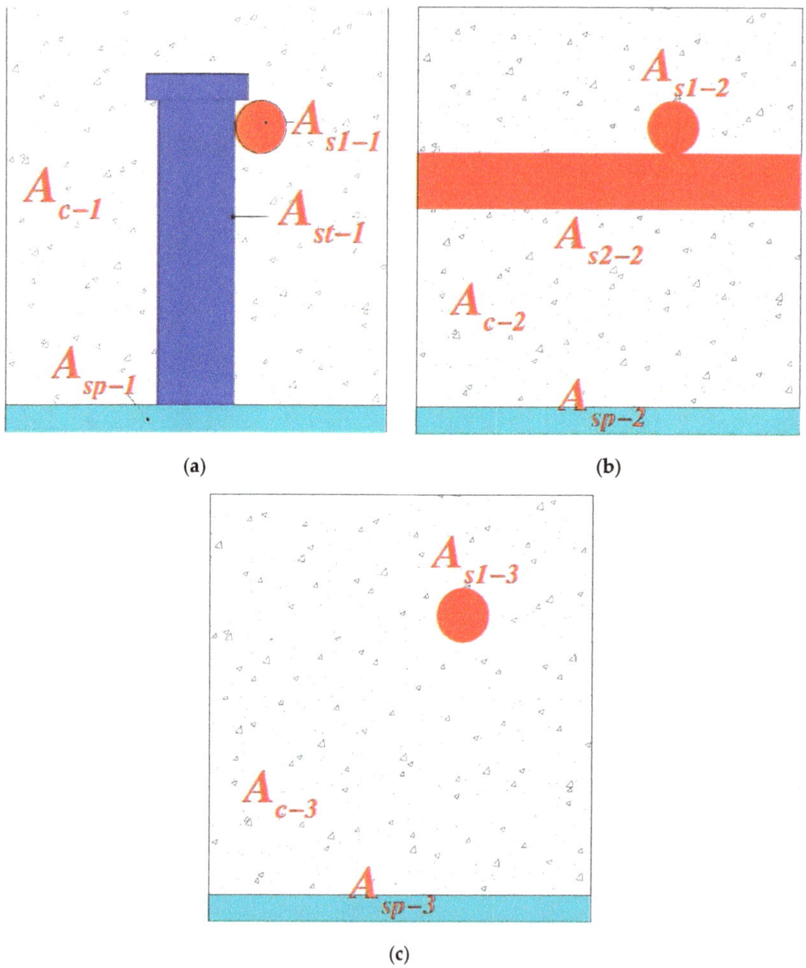

Figure 6. Three typical sections in restrained concrete: (**a**) Type 1 cross-section; (**b**) Type 2 cross-section; (**c**) Type 3 cross-section.

In the length direction (x-direction) of the specimen, the concrete is subjected to capillary pore stresses and confining tensile stresses. In this study, the primarily considered restraining effect is that of steel plates and pins in the length direction (x), so only capillary stresses act on the concrete units in the width direction (y) and height direction (z). According to the theory on the mechanics of materials, the strain in the x direction of concrete in sections 1, 2 and 3 can be expressed as

$$\varepsilon_{c-i} = \frac{1}{E_{SC}}\left[\sigma_x - \mu\left(\sigma_y + \sigma_z\right)\right] = \frac{1}{E_{SC}}\left[\left(\sigma_{ca} - \sigma_{t-i}\right) - \mu(\sigma_{ca} + \sigma_{ca})\right] = \frac{1-2\mu}{E_{SC}}\sigma_{ca} - \frac{1}{E_{SC}}\sigma_{t-i} \quad (2)$$

where ε_{c-i} is the strain of the concrete in section i in the restrained member and i = 1, 2, 3 in this study; E_{sc} is the modulus of elasticity of the concrete with respect to the capillary pore stresses; σ_{t-i} is the restrained tensile stresses on the concrete in section i, i = 1, 2, 3; and μ is the Poisson's ratio.

The total displacement of the whole concrete specimen under shrinkage stress (ΔL) can be expressed as

$$\Delta L = \varepsilon_{c-1}nd + \varepsilon_{c-2}ND + \varepsilon_{c-3}(L - nd - ND) \quad (3)$$

The total strain of the whole constrained model can then be expressed as

$$\varepsilon_{c-total} = \frac{\Delta L}{L} = \frac{\varepsilon_{c-1}nd + \varepsilon_{c-2}ND + \varepsilon_{c-3}(L - nd - ND)}{L} \quad (4)$$

where $\varepsilon_{c-total}$ is the total strain in the concrete restrained by the steel plate, pins, and reinforcement.

An arbitrary length (l) of confined concrete containing only one section on the model is considered. The concrete, steel plates, pins, and horizontal reinforcement in this section have the same displacement under capillary pore stresses, as required by the deformation coordination:

$$\delta_{c-1} = \delta_{sp-1} = \delta_{st-1} = \delta_{s1-1} \quad (5)$$

where δ_{c-1}, δ_{sp-1}, δ_{st-1}, and δ_{s1-1} are the displacements of concrete, steel plates, pins, and longitudinal reinforcement under capillary pore stresses for type 1 concrete with a length of l.

It can be inferred that

$$\frac{\delta_{c-1}}{l} = \frac{\delta_{sp-1}}{l} = \frac{\delta_{st-1}}{l} = \frac{\delta_{s1-1}}{l} = \delta_{c-1} = \delta_{sp-1} = \delta_{st-1} = \delta_{s1-1} \quad (6)$$

A constrained concrete containing only 2 sections of any length (l) on the model is considered. From the deformation coordination requirement, the following can be derived:

$$\varepsilon_{c-2} = \varepsilon_{sp-2} = \varepsilon_{st-2} = \varepsilon_{s2-2} \quad (7)$$

A constrained concrete containing only three sections of any length (l) on the model is considered. From the deformation coordination requirement, the following can be derived:

$$\delta_{c-2} = \delta_{sp-3} = \delta_{s1-3} \quad (8)$$

Organizing the above equation yields

$$\begin{aligned}
\varepsilon_{c-1} &= \frac{2\gamma(1-2\mu)}{r} \cdot \frac{1}{E_{SC} + \frac{A_{sp-1}}{A_{c-1}}E_{sp} + \frac{A_{st-1}}{A_{c-1}}E_{st} + \frac{A_{s1-1}}{A_{c-1}}E_S} \\
\varepsilon_{c-2} &= \frac{2\gamma(1-2\mu)}{r} \cdot \frac{1}{E_{SC} + \frac{A_{sp-2}}{A_{c-2}}E_{sp} + \frac{A_{s1-2}}{A_{c-2}}E_S + \frac{A_{s2-2}}{A_{c-2}}E_S} \\
\varepsilon_{c-2} &= \frac{2\gamma(1-2\mu)}{r} \cdot \frac{1}{E_{SC} + \frac{A_{sp-3}}{A_{c-3}}E_{sp} + \frac{A_{s1-3}}{A_{c-3}}E_S}
\end{aligned} \quad (9)$$

Substituting Equation (9) into Equation (4) yields

$$\varepsilon_{c-total} = \frac{2\gamma(1-2\mu)}{r}\left[\begin{array}{l} \frac{nd}{L}\cdot\frac{1}{E_{SC}+\frac{A_{sp-1}}{A_{c-1}}E_{sp}+\frac{A_{st-1}}{A_{c-1}}E_{st}+\frac{A_{s1-1}}{A_{c-1}}E_S} \\ +\frac{ND}{L}\cdot\frac{1}{E_{SC}+\frac{A_{sp-2}}{A_{c-2}}E_{sp}+\frac{A_{s1-2}}{A_{c-2}}E_S+\frac{A_{s2-1}}{A_{c-2}}E_S} \\ +\left(\frac{L-\mathrm{nd}-ND}{L}\right)\cdot\frac{1}{E_{SC}+\frac{A_{sp-3}}{A_{c-3}}E_{sp}+\frac{A_{s1-3}}{A_{c-3}}E_S} \end{array}\right] \tag{10}$$

Letting $\rho = nd/L$ and $\rho' = ND/L$, the equation for the shrinkage strain of concrete restrained by steel plates, pins, and reinforcement can be derived as

$$\varepsilon_{c-total} = \frac{2\gamma(1-2\mu)}{r}\left[\begin{array}{l} \rho\cdot\frac{1}{E_{SC}+\frac{A_{sp-1}}{A_{c-1}}E_{sp}+\frac{A_{st-1}}{A_{c-1}}E_{st}+\frac{A_{s1-1}}{A_{c-1}}E_S} \\ +\rho'\cdot\frac{1}{E_{SC}+\frac{A_{sp-2}}{A_{c-2}}E_{sp}+\frac{A_{s1-2}}{A_{c-2}}E_S+\frac{A_{s2-1}}{A_{c-2}}E_S} \\ +(1-\rho-\rho')\cdot\frac{1}{E_{SC}+\frac{A_{sp-3}}{A_{c-3}}E_{sp}+\frac{A_{s1-3}}{A_{c-3}}E_S} \end{array}\right] \tag{11}$$

It can be reordered as follows:

$$\frac{1}{E_{SC}+\frac{A_{sp-1}}{A_{c-1}}E_{sp}+\frac{A_{st-1}}{A_{c-1}}E_{st}+\frac{A_{s1-1}}{A_{c-1}}E_S} = A$$

$$\frac{1}{E_{SC}+\frac{A_{sp-2}}{A_{c-2}}E_{sp}+\frac{A_{s1-2}}{A_{c-2}}E_S+\frac{A_{s2-1}}{A_{c-2}}E_S} = B$$

$$\frac{1}{E_{SC}+\frac{A_{sp-3}}{A_{c-3}}E_{sp}+\frac{A_{s1-3}}{A_{c-3}}E_S} = C$$

Finally, the following is obtained:

$$\varepsilon_{c-total} = \frac{2\gamma(1-2\mu)}{r}\left(\frac{\rho}{A}+\frac{\rho'}{B}+\frac{1-\rho-\rho'}{C}\right) \tag{12}$$

4. Hole Structures in Complex Internally Confined Concrete

4.1. Confinement Factor of Concrete by Steel Plates, Pins, and Reinforcement Bars

From Equation (11), the average combined force $\overline{\sigma}_{c-total}$ exerted on the concrete hole wall throughout the specimen with the joint restraint of the steel plate, studs, and reinforcement can be expressed as

$$\overline{\sigma}_{c-total} = E_{SC}\frac{2\gamma(1-2\mu)}{r}\left[\begin{array}{l}\rho\cdot\dfrac{1}{E_{SC}+\dfrac{A_{sp-1}}{A_{c-1}}E_{sp}+\dfrac{A_{st-1}}{A_{c-1}}E_{st}+\dfrac{A_{s1-1}}{A_{c-1}}E_{S}}\\+\rho'\cdot\dfrac{1}{E_{SC}+\dfrac{A_{sp-2}}{A_{c-2}}E_{sp}+\dfrac{A_{s1-2}}{A_{c-2}}E_{S}+\dfrac{A_{s2-1}}{A_{c-2}}E_{S}}\\+(1-\rho-\rho')\cdot\dfrac{1}{E_{SC}+\dfrac{A_{sp-3}}{A_{c-3}}E_{sp}+\dfrac{A_{s1-3}}{A_{c-3}}E_{S}}\end{array}\right] \quad (13)$$

To measure the restraining effect of steel plates, bolts, and reinforcement on concrete, a parameter λ called the restraining coefficient is proposed:

$$\lambda = \frac{1}{(1-2\mu)\left[\begin{array}{l}\rho\cdot\dfrac{1}{E_{SC}+\dfrac{A_{sp-1}}{A_{c-1}}E_{sp}+\dfrac{A_{st-1}}{A_{c-1}}E_{st}+\dfrac{A_{s1-1}}{A_{c-1}}E_{S}}\\+\rho'\cdot\dfrac{1}{E_{SC}+\dfrac{A_{sp-2}}{A_{c-2}}E_{sp}+\dfrac{A_{s1-2}}{A_{c-2}}E_{S}+\dfrac{A_{s2-1}}{A_{c-2}}E_{S}}\\+(1-\rho-\rho')\cdot\dfrac{1}{E_{SC}+\dfrac{A_{sp-3}}{A_{c-3}}E_{sp}+\dfrac{A_{s1-3}}{A_{c-3}}E_{S}}\end{array}\right]} \quad (14)$$

Equation (14) shows that the constraint factor will increase when the thickness of the steel plate is increased, increasing the diameter of the reinforcement and the diameter and height of the studs while decreasing the spacing of the studs at the same age (constant concrete E_{sc}).

Substituting λ into Equation (13) yields the following:

$$\overline{\sigma}_{c-total} = \frac{1}{\lambda}E_{SC}\frac{2\gamma}{r} \quad (15)$$

The confinement coefficients at 3, 7, and 28 d for each specimen in this study are presented in Table 4. Among these, the elastic modulus E_{sc} of concrete relative to the capillary stress can be converted from the static compressive elastic modulus E_c of concrete. The ratio of E_c to E_{SC} varies between 2.5 and 3.5 as the concrete strength increases. In this study, the ratio was set to 3.2. The static compressive modulus of elasticity of concrete at various ages is listed in Table 3, and the Poisson's ratio of concrete is taken to be 0.2 as per the relevant specification. Accordingly, the constraints on the concrete in each specimen are as follows: S5 < S1 < S4 < S6 < S3 < S2.

Table 4. λ of constraint coefficient of steel plate, stud, and reinforcement on concrete.

No.	Constraint Factor ($\times 10^4$ MPa)		
	3 Days	**7 Days**	**28 Days**
S1	3.75	3.87	4.04
S2	4.15	4.27	4.44
S3	3.96	4.08	4.26
S4	3.79	3.91	4.09
S5	3.72	3.84	4.01
S6	3.85	3.97	4.14

4.2. Influence of Steel Plates, Pins, and Reinforcement on the Structure of Concrete Holes

Figure 7 shows the variation in the internal pore structure parameters (the most cumulative pore size, porosity, average pore size, and median pore size) of concrete with a constraint coefficient. Noticeably, the pore structure of concrete changes with the configuration of steel plates, bolts, and steel bars. Testing the pore structure of concrete with different bolt configurations reveals that increasing the diameter of the bolt will increase the pore structure of concrete when the height and spacing of the bolt remain unchanged. For example, when the height of the stud is maintained at 80 mm and the spacing between the studs is maintained at 200 mm, the S1 specimen with a stud diameter of 16 mm has the highest cumulative pore size, porosity, average pore size, and median pore size of 37.0 nm, 12.1%, 35.9 nm, and 51.9 nm at 3 d aging, respectively. Compared with the S4 specimen with a 19 mm diameter stud, the same parameters were reduced by 3.5 nm, 1.5%, 4.7 nm, and 9.0 nm, respectively.

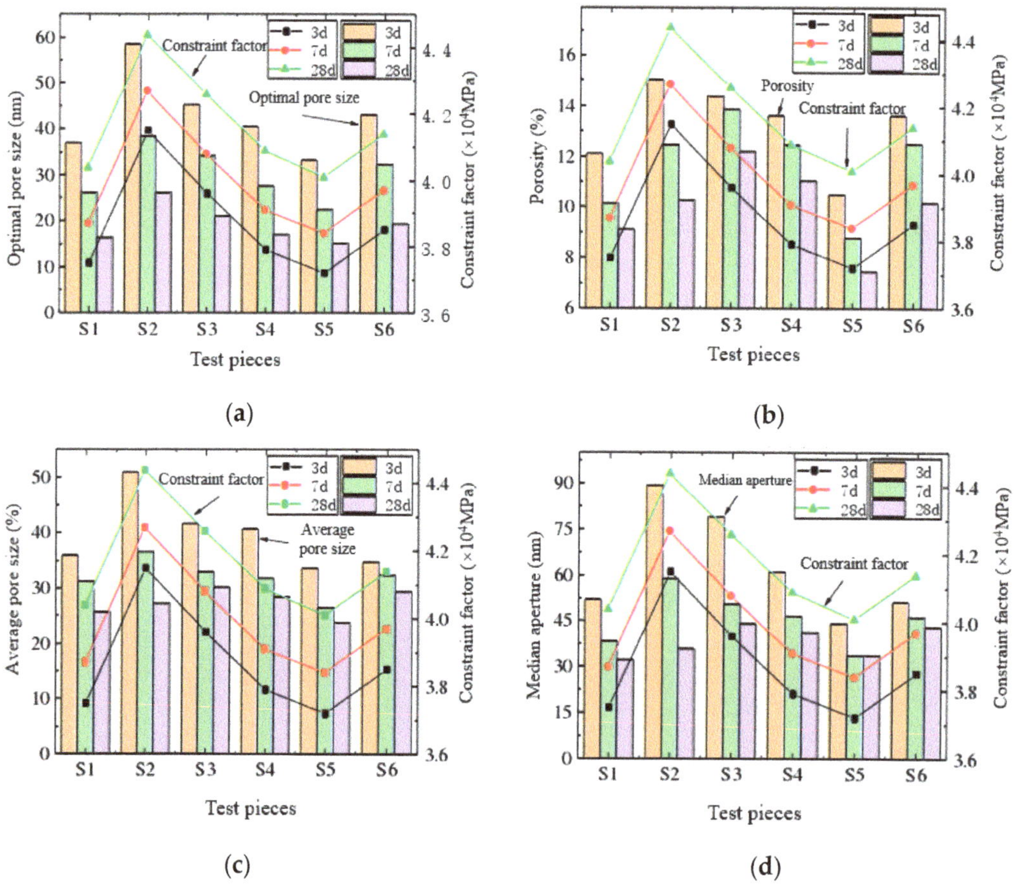

Figure 7. Variation in concrete pore structure with constraint coefficient: (**a**) optimal pore size; (**b**) porosity; (**c**) average pore size; (**d**) median aperture.

Reducing the spacing of studs expands the pore structure of concrete when the thickness of the steel plate, diameter of the reinforcement, and height and diameter of the studs are kept constant. For example, when the heights of the studs were both 80 mm and the diameters of the studs were both 16 mm, the highest cumulative pore size, porosity, mean pore size, and median pore size was observed in the S3 specimen with a 100 mm stud spacing, corresponding to 45.3 nm, 14.4%, 41.6 nm, and 79.0 nm, respectively, at 3 d. These were higher than those of the same parameters of the S1 specimen with 200 mm stud spacing, which were 8.3 nm, 2.3%, 5.7 nm and 27.1 nm, respectively.

When the plate thickness, rebar diameter, spigot diameter, and spigot spacing were kept constant, increasing the spigot height expanded the pore structure of the concrete. For example, when the spacings of the studs were all 200 mm and the diameters of the studs were all 16 mm, the highest cumulative pore size, porosity, average pore size, and median pore size were observed in the S5 specimen with a stud height of 60 mm at 3 d, corresponding to 33.2 nm, 10.5%, 33.6 nm, and 44.2 nm, respectively. These represented a decrease compared with the same parameters in the S1 specimen with a stud height of 80 mm by 3.8 nm, 1.6%, 2.2 nm, and 7.7 nm, respectively.

The effect of steel reinforcement on the pore structure of concrete can be determined by comparing the pore structure of these specimens when the thickness of the steel plate is the same as the relevant parameters of the studs (diameter, height, and spacing). Testing the pore structure of concrete with different reinforcement configurations reveals that increasing the diameter of the reinforcement will expand the pore structure. For example, when the thickness of the steel plate was 10 mm, the heights of the studs were all 80 mm, the spacing of the studs were all 200 mm, and the diameters of the studs were all 16 mm. Moreover, the highest cumulative pore size, porosity, average pore size, and median pore size were observed in specimen S6 with a 12 mm reinforcement diameter at 7 d corresponding to 32.4 nm, 12.5%, 32.4 nm, and 46.3 nm, respectively, which were lower than those of a similar specimen S1 by 6.1 nm, 2.3%, 1.2 nm, and 8.1 nm.

The effect of reinforcement on the pore structure of concrete can be investigated by comparing the pore structure of these specimens when the reinforcement parameters are the same as the relevant parameters of the studs (diameter, height, and spacing). By testing the pore structure of concrete with different steel plate configurations, we observed that increasing the thickness of the steel plate expands the pore structure of concrete. For example, when the diameter of the reinforcement is 10 mm, the height of the studs is 80 mm, the spacing of the studs is 200 mm, and the diameter of the studs is 16 mm. Moreover, the highest cumulative pore size, porosity, average pore size, and median pore size were observed in the S2 specimens with a steel plate thickness of 12 mm after 3 d, corresponding to 58.5 nm, 15.0%, 50.9 nm, and 89.4 nm, which were higher than those of S1 specimens equipped with steel plates with a thickness of 10 mm, whose values were 21.5 nm, 2.9%, 15.0 nm, and 37.5 nm, respectively.

As shown in Equation (15), the combined force exerted on the concrete pore walls in the restrained specimens decreased with the increase in the restraining coefficient. Therefore, the stronger the restraint of the concrete by steel plate, studs, and reinforcement, the smaller is the combined force on the concrete pore walls and the lower the shrinkage in the concrete pore structure caused by the combined force, leading to an increase in the pore structure parameters of the concrete. This is consistent with experimental findings; for example, with the same diameter of reinforcement and the same parameters related to bolts, the constraint coefficient of the S1 specimen is less than that of the S2 specimen. Moreover, at 7 d, the highest cumulative pore size, porosity, mean pore size, and median pore size were observed in the S2 specimen, corresponding to 58.5 nm, 15.0%, 50.9 nm, and 89.4 nm, which is higher than that of the same parameter of the S1 member at the same age, whose values were 21.5 nm, 2.9%, 15.0 nm, and 37.5 nm, respectively. This can be interpreted as an increase in the confinement coefficient of S2 due to the increase in the thickness of the steel plate, and the increase in the confinement coefficient leads to a decrease in the combined force applied to the concrete pore walls. Further, the pore structure parameter of the S2 specimen increased in this case. The changes in the pore structure parameters of the concrete due to changes in the diameter of the reinforcement and the parameters of the studs were also observed in the tests.

5. Shrinkage of Complex Internally Confined Concrete

5.1. Effect of Constraint Factor on Concrete Shrinkage

The relationship between the shrinkage deformation of concrete and the constraint coefficient at different ages is shown in Figure 8. Noticeably, for the concrete specimens constrained by the steel plate, studs, and reinforcement, the shrinkage strain is affected by the thickness of the steel plate, diameter of the reinforcement, and the parameters related to the studs when the studs have the same diameter, height, and spacing, and the diameter of the reinforcement is kept the same. When the materials used for the steel plate are the same, the thicker the plate is, the smaller the shrinkage of concrete, as shown in Equations (11) and (15). At the same age, the greater the thickness of the steel plate is, the greater the constraint factor. An increase in the constraint factor leads to a decrease in the combined force acting on the concrete, which results in a decrease in the shrinkage of the concrete. In this study, specimen S2, with a steel plate thickness of 12 mm, has constraint coefficients of 4.15×10^4 MPa, 4.27×10^4 MPa, and 4.44×10^4 MPa at the ages of 3, 7, and 28 d, respectively, which were higher than the constraint coefficients of specimen S1 with a steel plate thickness of 10 mm at the same ages, corresponding to 0.4×10^4 MPa, 0.4×10^4 MPa, and 0.4×10^4 MPa, respectively. The shrinkage values of the S2 specimens at 3, 7, and 28 d were 66 με, 97 με, and 149 με, which were 46.4%, 39.8%, and 35.8% lower than the shrinkage values of the S1 specimens, respectively.

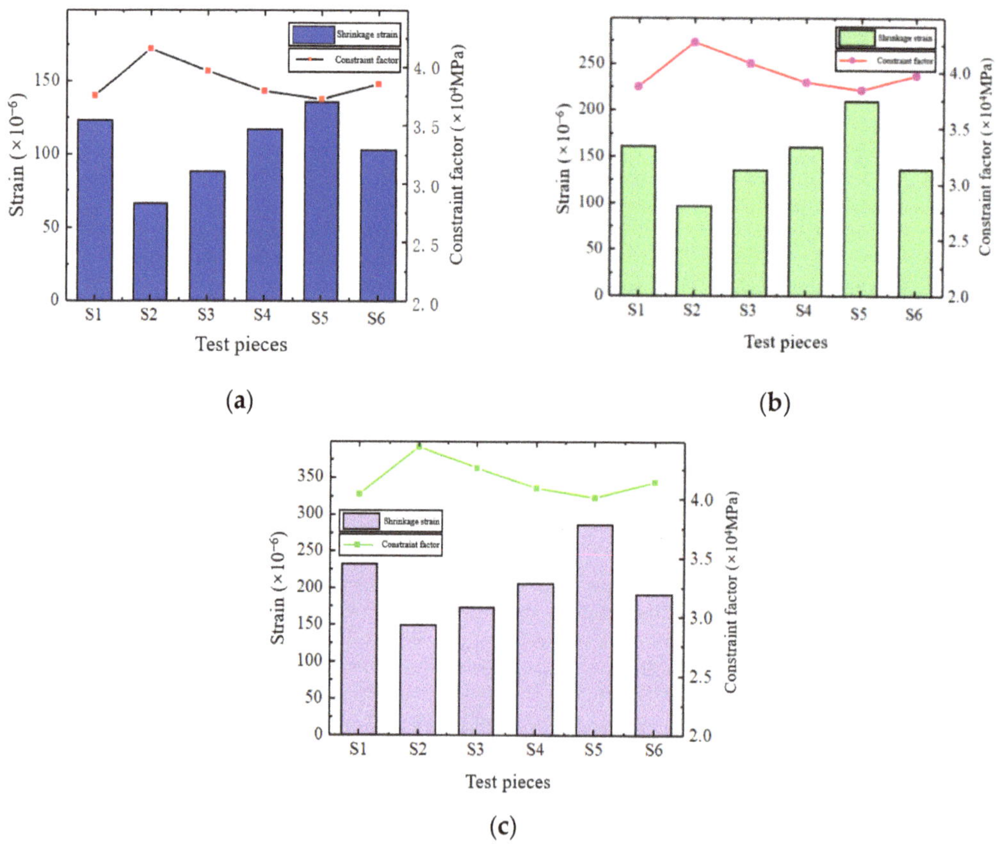

Figure 8. Diagram of shrinkage of concrete with constraint coefficient: (**a**) 3 days of age; (**b**) 7 days of age; (**c**) 28 days of age.

Figure 8 shows that, when the thickness of the steel plate and the diameter of the reinforcement bars are kept constant, the shrinkage values of the concrete are related to the relevant parameters of the studs (diameter, height, and spacing). Moreover, when the height of the studs is kept at 80 mm, and the spacing of the studs is kept at 200 mm, comparing the shrinkage data of the concrete from S1 to S4 revealed that the increase in the diameter of the studs increased the restraining capacity of the studs on the concrete (constraint factor rises), resulting in a decrease in concrete shrinkage (see Figure 8). This result is also observed in Equations (11) and (15). At the same age, the larger the diameter of the studs, the greater the constraint factor. An increase in the constraint factor leads to a decrease in the combined force acting on the concrete, which results in a decrease in the shrinkage of the concrete. In this study, the constraint coefficients of specimen S4 at 3, 7, and 28 d were 3.79×10^4 MPa, 3.91×10^4 MPa, and 4.09×10^4 MPa, respectively, which were higher than those of specimen S1 at the same ages by 0.04×10^4 MPa, 0.04×10^4 MPa, and 0.05×10^4 MPa, respectively. Regarding the constraint coefficients of specimen S4 at ages of 3, 7, and 28 d, the shrinkage values were 117 με, 160 με, and 206 με, respectively, which were lower than those of specimen S1 by 6.33 με, 0.67 με, and 25 με.

When the thickness of the steel plate, diameter of the reinforcement, diameter of the studs, and the spacing of the studs were kept constant, increasing the height of the studs increased the restraining ability of the studs on the concrete (the restraining coefficient rises), as revealed by comparing the concrete shrinkage data of S1 and S5, which led to a decrease in the shrinkage of the concrete (see Figure 8). At the same age, the greater the height of the stud is, the greater the constraint factor. The increase in the constraint factor leads to a decrease in the combined force acting on the concrete, which results in a decrease in the shrinkage of the concrete. In this study, the constraint coefficients of specimen S1 at the ages of 3, 7, and 28 d were 3.75×10^4 MPa, 3.87×10^4 MPa, and 4.04×10^4 MPa, respectively, which were higher than those of specimen S5 at the same ages by 0.03×10^4 MPa, 0.03×10^4 MPa, and 0.03×10^4 MPa, respectively. The constraint coefficients of specimen S1 at ages of 3, 7, and 28 d for the shrinkage values were 124 με, 161 με, and 232 με, respectively, which were 12 με, 48 με, and 53 με less than those of the S5 specimens.

When the plate thickness, rebar diameter, spigot diameter, and spigot height were kept constant, decreasing the spacing of the spigots increased the ability of the spigot to confine the concrete (confinement coefficient rises), which leads to a decrease in the shrinkage of the concrete, as revealed by comparing the concrete shrinkage data of S1 and S3. The same results are obtained from Equations (11) and (15). At the same age, the smaller the spacing of the studs is, the greater the constraint factor. The increase in the constraint factor leads to a decrease in the combined force acting on the concrete, which results in a decrease in the shrinkage of the concrete. In this study, the constraint coefficients of specimen S3 at the ages of 3, 7, and 28 d were 3.96×10^4 MPa, 4.08×10^4 MPa, and 4.26×10^4 MPa, respectively, which were higher than those of specimen S1 at the same ages by 0.21×10^4 MPa, 0.21×10^4 MPa, and 0.22×10^4 MPa, respectively. For the constraint coefficients of specimen S3 at ages of 3, 7, and 28 d, the shrinkage values were 88 με, 136 με, and 174 με, respectively, which were 28.8%, 15.6%, and 25.1% lower than those of the S4 specimens.

When the thickness of the steel plate and the parameters related to the bolts (height, spacing, and diameter) were kept constant, the shrinkage of the concrete was affected by the diameter of the reinforcement bars, as revealed by comparing the concrete shrinkage data of S1 and S6. Moreover, when the diameter of the reinforcement bars is larger, the shrinkage of the concrete is smaller (see Figure 8). This result is corroborated by Equations (11) and (15). At the same age, the larger the diameter of the reinforcement, the larger is the constraint factor. An increase in the constraint coefficient leads to a decrease in the combined force acting on the concrete, which results in a decrease in the shrinkage of the concrete. In this study, the constraint coefficients of specimen S6 at the ages of 3, 7, and 28 d were 3.85×10^4 MPa, 3.97×10^4 MPa, and 4.14×10^4 MPa, respectively, which were higher than those of specimen S1 at the same ages by 0.1×10^4 MPa, 0.1×10^4 MPa, and 0.1×10^4 MPa, respectively. Moreover, the constraint coefficients of specimen S6 at 3, 7, 28 d corresponded to shrinkage values of S6 103 με, 136 με, and 191 με, respectively, which were 16.7%, 15.4%, and 17.7% less than those of the S1 specimen.

5.2. Effect of Constraint Coefficient on Capillary Pore Stresses

Capillary pore stress is the main driver of shrinkage deformation in concrete. Table 5 and Figure 9 present the capillary pore stresses for each specimen at 3, 7, and 28 d. The test results reveal that the capillary pore stresses are affected by the thickness of steel plates, the diameter of reinforcement, and the relevant parameters of studs in concrete specimens restrained by steel plates, studs, and reinforcement because changing the respective parameters at the same age changes the restraining coefficients of the concrete. Moreover, greater restraining coefficients of the concrete result in the pore structural parameters (the maximum allowable pore diameters, porosity, average pore size, median pore size, and others) of the restraining specimen increasing, which, according to Equation (1), leads to a reduction in capillary pore stress in concrete.

Table 5. Capillary stress of concrete at different ages.

No.	Capillary Stress (MPa)		
	3 Days	7 Days	28 Days
S1	0.99	1.44	2.44
S2	0.56	0.89	1.38
S3	0.76	1.05	1.80
S4	0.89	1.36	2.31
S5	1.11	1.71	2.65
S6	0.82	1.14	2.02

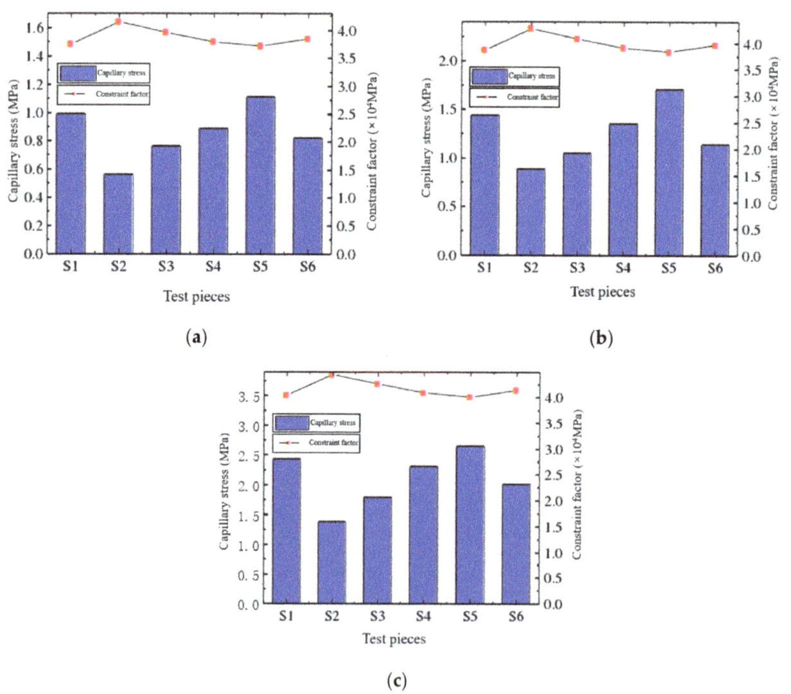

Figure 9. Relationship between capillary stress and constraint coefficient of concrete at different ages: (**a**) 3 days of age; (**b**) 7 days of age; (**c**) 28 days of age.

Table 5 and Figure 9 reveal that the thicker the steel plate, the lower the capillary pore stresses when the diameter, height, and spacing of the studs are kept constant and the diameters of the reinforcement and the material used for the steel plate are the same. This is because the constraint factor increases when the thickness of the steel plate is greater at the same age. The increase in the confinement factor causes the pore structure of the confined specimen to expand, leading to a decrease in the capillary pore stress in concrete (see Figure 9). In this study, the constraint coefficients of specimen S1 and specimen S2 were sequentially increased, and a sequential decrease in their capillary pore stresses was observed in the tests. For example, at the ages of 3, 7, and 28 d, the capillary pore stresses of specimen S2 were 0.56 MPa, 0.89 MPa, and 1.40 MPa, respectively, which were 0.42, 0.55, and 1.01 lower than the capillary pore stresses of specimen S1 at the same ages.

The analysis in Figure 9 reveals that, when the steel plate thickness and the diameter of the reinforcement bars remain unchanged, the concrete capillary pore stress is related to the stud parameters (diameter, height, and spacing). Moreover, when the height of the studs was kept at 80 mm and the spacing of the studs at 200 mm, the capillary pore stresses of the concrete in contrasting test specimens S1 and S4 reveal that the increase in the diameter of the studs increased the capacity (constraint coefficient increases) of the studs to restrain the concrete (constraint coefficient rises), decreasing the capillary pore stress of the concrete (see Figure 9). For example, the confinement coefficients of specimen S4 at the ages of 3, 7, and 28 d were 3.79×10^4 MPa, 3.91×10^4 MPa, and 4.09×10^4 MPa, respectively, which were higher than those of specimen S2 at the same ages by 0.04×10^4 MPa, 0.04×10^4 MPa, and 0.05×10^4 MPa, and the corresponding capillary pore stresses were reduced from 0.99 MPa, 1.44 MPa, and 2.44 MPa in specimen S1 to 0.89 MPa, 1.36 MPa, and 2.31 MPa in specimen S4.

When the thickness of the steel plate, diameter of the reinforcement, diameter of the studs, and spacing of the studs were kept constant, increasing the height of the studs increased the ability of the studs to restrain the concrete (the restraining coefficient rises), as revealed by the comparisons of the concrete capillary pore stress data of specimens S1 and S5 in Figure 9, which leads to a decrease in the capillary pore stresses of the concrete (see Figure 9). For example, at the ages of 3, 7, and 28 d, the confinement coefficients of the S1 specimens with a stud height of 80 mm were 3.75×10^4 MPa, 3.87×10^4 MPa, and 4.04×10^4 MPa, respectively, which were higher than the confinement coefficients of the S5 specimens configured with a stud height of 60 mm at the same ages, which corresponded to 0.03×10^4 MPa, 0.03×10^4 MPa, and 0.03×10^4 MPa, respectively. Moreover, the capillary pore stresses of the S1 specimen at 3, 7, and 28 d were 0.99 MPa, 1.44 MPa, and 2.44 MPa, respectively, which were, respectively, 0.12 MPa, 0.27 MPa, and 0.21 MPa lower than those of the S5 specimen.

When the thickness of the steel plate, diameter of the reinforcement, diameter of the studs, and height of the studs were kept constant, comparing the concrete capillary pore stress data of specimens S1 and S3 in Figure 9 reveals that the reduction in the spacing of the studs increases the ability of the studs to restrain the concrete (the restraining coefficient rises), which leads to a decrease in the capillary pore stresses of the concrete (see Figure 9). For example, at the ages of 3, 7, and 28 d, the constraint coefficients of the S3 specimens configured with a stud spacing of 100 mm were 3.96×10^4 MPa, 4.08×10^4 MPa, and 4.26×10^4 MPa, respectively, which were higher than the constraint coefficients of S1 specimens configured with a stud spacing of 200 mm at the same ages, corresponding to 0.21×10^4 MPa, 0.21×10^4 MPa, and 0.22×10^4 MPa, respectively. Moreover, the capillary pore stresses of the S3 specimens at 3, 7, and 28 d were 0.76 MPa, 1.05 MPa, and 1.80 MPa, respectively, which in turn were 0.22 MPa, 0.40 MPa, and 0.64 MPa lower than the capillary pore stresses of the S1 specimens.

When the thickness of the steel plate and the relevant parameters (height, spacing, and diameter) of the bolts were kept constant, the capillary pore stress of the concrete was affected by the diameter of the reinforcement bars, as revealed by the comparison of specimens S1 and S6 in Figure 9. When the diameter of the reinforcement bars was larger, the capillary pore stress of the concrete was smaller. This is because, at the same age, the larger the diameter of the reinforcement bar, the larger the confinement coefficient, which expands the pore structure of the confined specimen, as shown in Equation (1), leading to a decrease in the capillary pore stress in concrete. In this study, the constraint coefficients of specimens S1 and S6 increased due to the larger diameter of the reinforcement bars. Accordingly, a sequential decrease in their capillary pore stresses was observed in the tests. For example, the confinement coefficients of specimen S6 at 3, 7, and 28 d were 3.85×10^4 MPa, 3.97×10^4 MPa, and 4.09×10^4 MPa, respectively, which were higher than the confinement coefficients of specimen S1 at the same ages by 0.1×10^4 MPa, 0.1×10^4 MPa, and 0.1×10^4 MPa, respectively, and those of the S6 specimen at the ages of 3, 7, and 28 d. The capillary pore stresses of the S6 specimen at 3, 7, and 28 d of age were 0.82 MPa, 1.14 MPa, and 2.02 MPa, respectively, which were 0.16 MPa, 0.30 MPa, and 0.42 MPa lower than those of the S1 specimen.

5.3. Effect of Capillary Pore Stress on Concrete Shrinkage

The relationship between the capillary pore stress and concrete shrinkage of specimens at different ages is shown in Figure 10. Noticeably, the greater the capillary pore stress, the greater the shrinkage deformation of concrete. Meanwhile, according to the previous analysis, the concrete capillary pore stress decreases with the increase in the constraint coefficient, which suggests that the constraining effect of the steel plate, studs, and rebar on the shrinkage of the concrete is not only due to the increase in the constraining tensile stress because of the coordination of the deformation between the concrete, the steel plate, the studs, and the steel rebar but also the decrease in the capillary pore stress in the concrete due to the presence of the steel plate, studs, and rebar; the principle of this action is shown in Figure 11.

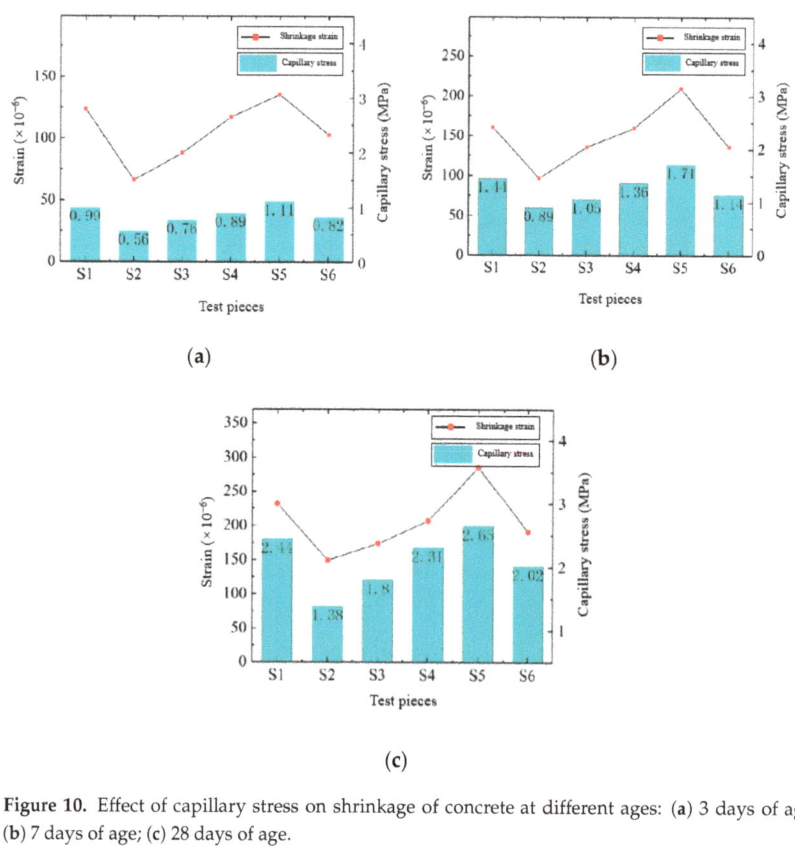

Figure 10. Effect of capillary stress on shrinkage of concrete at different ages: (**a**) 3 days of age; (**b**) 7 days of age; (**c**) 28 days of age.

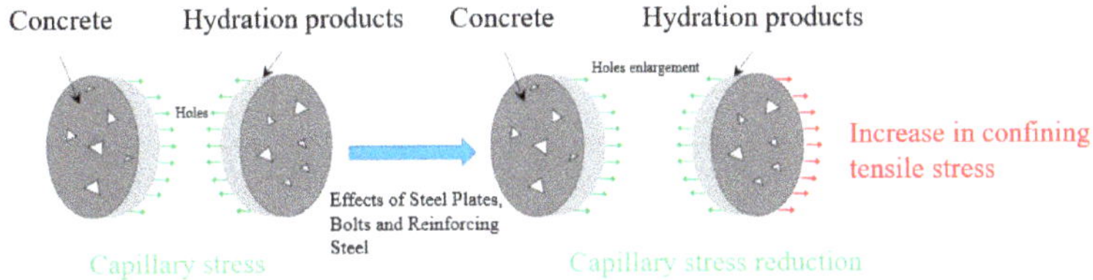

Figure 11. Schematic of the effect of steel plate, stud, and steel bar on the shrinkage of concrete.

5.4. Shrinkage Prediction of Complex Internally Confined Concrete

The prediction of shrinkage in restrained concrete is important for assessing the risk of concrete shrinkage cracking. In this paper, a prediction model for the shrinkage of concrete restrained by steel plates, spigots, and reinforcement is derived based on the capillary pore stress theory (Equation (11)). To verify the accuracy of this model, the shrinkage test values of various types of restrained concrete specimens at the ages of 3, 7, and 28 d were compared with the calculated values obtained through shrinkage modeling. The modulus of elasticity of steel plates and spigots was taken as 206,000 MPa, as suggested in GB 50017-2017 [29]. The maximum number of holes for each type of specimen at each age is given in Figure 7.

Table 6 and Figure 12 present the results of the comparison between the shrinkage test results and the predicted values derived from the shrinkage model calculation. Figure 12 shows that the calculated values are in good agreement with the measured values, and the average value of the prediction error was calculated to be less than 15%, which indicates that the constrained concrete shrinkage model proposed in this paper can effectively predict the shrinkage of concrete constrained by steel plates, pins, and reinforcement.

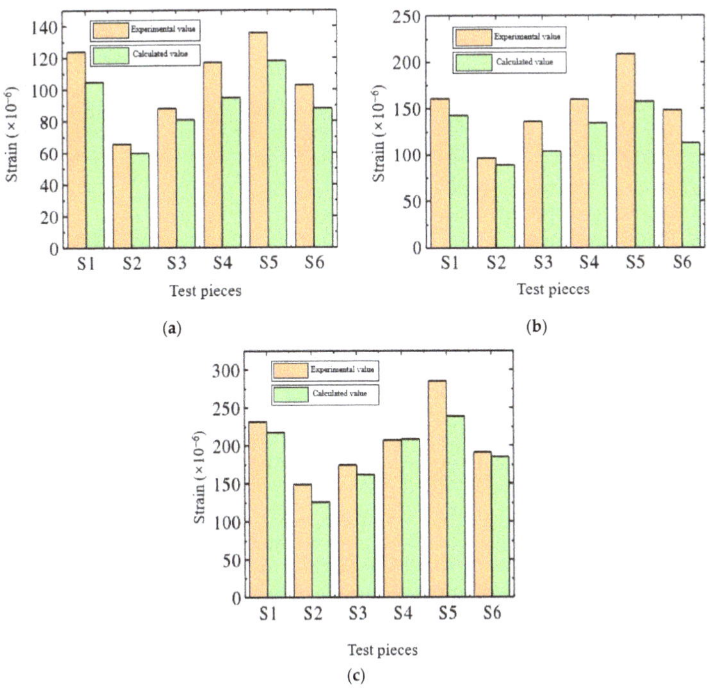

Figure 12. Comparison of test results and calculated values at different ages: (**a**) 3 days of age; (**b**) 7 days of age; (**c**) 28 days of age.

Table 6. Comparison of test results and calculated values.

No.	3 Days			7 Days			28 Days		
	Tested Values	Calculated Values	Relative Error	Tested Values	Calculated Values	Relative Error	Tested Values	Calculated Value	Relative Error
S1	124	105	15.0%	161	143	10.9%	232	220	5.3%
S2	66	60	9.5%	97	89	8.4%	149	125	16.3%
S3	88	81	7.7%	136	104	23.4%	174	162	6.8%
S4	117	95	19.1%	160	135	15.7%	207	208	0.9%
S5	136	118	13.2%	209	169	19.2%	285	239	16.4%
S6	103	88	14.9%	136	113	16.8%	191	182	4.7%

6. Conclusions

Through experiments and theoretical analysis, this study investigates the early shrinkage and pore structure of concrete under complex internal constraints. The following conclusions are drawn based on the research results:

1. The degree of the restraining of concrete by the steel plate, studs, and reinforcement is expressed by the restraining coefficient λ. The larger the restraining coefficient, the stronger the restraining effect. The constraint coefficient increases with the increase in the steel plate thickness, reinforcement diameter, stud diameter, and stud height, and increases with the decrease in stud spacing.
2. Steel plates, studs, and reinforcement have an important effect on concrete shrinkage, which decreases with the increase in the thickness of the steel plate, diameter of the reinforcement, diameter of the studs, and the height of the studs and increases with the increase in the stud spacing. This effect is not only due to the coordination of the deformation of the steel plate, studs, and reinforcement but also because the increase in the corresponding parameter increases the constraint coefficient. Moreover, the constraint factor will lead to a reduction in the capillary pore stress, which leads to a decrease in the shrinkage of the concrete.
3. Comparing the measured shrinkage strain values of restrained concrete at the ages of 3, 7, and 28 d with the predicted values of the shrinkage model revealed that the average error predicted during each age was below 15%. This proves that the shrinkage model is feasible for predicting the shrinkage strain of concrete under the joint constraints of steel plates, bolts, and steel bars.

The shrinkage of concrete, especially under constraint conditions, is a complex problem. Therefore, there are still some issues that need further research in this study. For example, this article only considers the shrinkage deformation of concrete in the one-dimensional length direction, while in practical engineering, the shrinkage deformation of steel plate concrete composite shear walls is three-dimensional, which is different from the actual situation. In future, it will be necessary to conduct further research on three-dimensional shrinkage deformation. In addition, although there are currently no reports on the impact of "size effect" on concrete shrinkage test results, whether there will be size effects and what impact size effects will have when many constraint conditions are added to concrete is also one of the research subjects worth examining in the future.

Author Contributions: F.Z. and H.J.: conceptualization, investigation, methodology, investigation, data curation, writing—original draft, writing—review and editing, formal analysis, L.H.: supervision, resources, project administration, funding acquisition writing—original draft, writing—review and editing, visualization. Y.H., Z.X., Z.Z., M.L., B.W. and X.Z.: writing—original draft, writing—review and editing. All authors have read and agreed to the published version of the manuscript.

Funding: This research received no external funding.

Data Availability Statement: Not applicable.

Acknowledgments: The authors greatly acknowledge the financial support for publication fees by the School of Civil Engineering, Chongqing University, China. Also, the authors are very grateful to Lin Chen for his contribution to this paper.

Conflicts of Interest: The authors declare no conflict of interest.

References

1. Guo, D.; Ma, Y.; Zhao, G.; Zhang, M.; Qiu, J. Seismic behavior of steel-plate encased concrete shear wall with paralleled circular CFST columns: Laboratory test and numerical simulation. *Structures* **2023**, *49*, 240–255. [CrossRef]
2. Wang, W.; Ren, Y.; Lu, Z.; Song, J.; Han, B.; Zhou, Y. Experimental study of the hysteretic behaviour of corrugated steel plate shear walls and steel plate reinforced concrete composite shear walls. *J. Constr. Steel Res.* **2019**, *160*, 136–152. [CrossRef]
3. Zhang, J.; Li, X.; Cao, W.; Yu, C. Seismic behavior of composite shear walls incorporating high-strength materials and CFST boundary elements. *Eng. Struct.* **2020**, *220*, 110994. [CrossRef]
4. Li, B.; Li, J.; Xu, W. *Dynamic Time Analysis of the Composite Shear Wall on the Grid-Tube-Type Double Steel Plate Wall with Infilled Concrete*; IOP Conference Series: Earth and Environmental Science; IOP Publishing: Bristol, UK, 2021; p. 012008.
5. Ge, W.; Zhang, Z.; Xu, W.; Ashour, A.; Jiang, H.; Sun, C.; Song, S.; Cao, D. Seismic response of grid tubular-double steel plate concrete composite shear walls and combined system subjected to low reversed cyclic loading. *Eng. Struct.* **2022**, *256*, 114028. [CrossRef]
6. Zhang, Z.; Xu, W.; Ge, W.; Abbas, E.M.A.; Jiang, H.; Wang, Y.; Yao, S. Seismic performance of grid tubular-double steel plate concrete composite shear wall. *J. Constr. Steel Res.* **2022**, *189*, 107077. [CrossRef]
7. Xiang, T.; Liu, J.; Lv, Z.; Wei, F.; Liu, Q.; Zhang, Y.; Ren, H.; Zhou, S.; Chen, D. The effect of silicon-based waterproof agent on the wettability of superhydrophobic concrete and enhanced corrosion resistance. *Constr. Build. Mater.* **2021**, *313*, 125482. [CrossRef]
8. Wang, L.; Yu, Z.; Liu, B.; Zhao, F.; Tang, S.; Jin, M. Effects of Fly Ash Dosage on Shrinkage, Crack Resistance and Fractal Characteristics of Face Slab Concrete. *Fractal Fract.* **2022**, *6*, 335. [CrossRef]
9. Safiuddin, M.; Kaish, A.B.M.A.; Woon, C.-O.; Raman, S.N. Early-Age Cracking in Concrete: Causes, Consequences, Remedial Measures, and Recommendations. *Appl. Sci.* **2018**, *8*, 1730. [CrossRef]
10. Shen, D.; Kang, J.; Liu, C.; Li, M.; Wei, Y.; Zhou, L. Effect of temperature rise inhibitor on early-age behavior and cracking resistance of high strength concrete under uniaxial restrained condition. *J. Build. Eng.* **2023**, *64*, 105588. [CrossRef]
11. Ali, B.; Hawreen, A.; Ben Kahla, N.; Talha Amir, M.; Azab, M.; Raza, A. A critical review on the utilization of coir (coconut fiber) in cementitious materials. *Constr. Build. Mater.* **2022**, *351*, 128957. [CrossRef]
12. Shang, M.; Qiao, H.; He, Z.; Zhang, Y.; Liu, X.; Zhang, Y.; Wang, J. Study on the shrinkage mechanism and prediction model of high-strength specified density concrete based on the internal curing micro-pump effect of MSW incineration tailings. *Constr. Build. Mater.* **2023**, *372*, 130804. [CrossRef]
13. Shen, D.; Li, C.; Liu, C.; Li, M.; Kang, J. Experimental investigations on early-age tensile creep of internally cured high strength concrete under different initial stress/strength ratios. *Constr. Build. Mater.* **2020**, *265*, 120313. [CrossRef]
14. Yoo, D.-Y.; Banthia, N.; Yoon, Y.-S. Effectiveness of shrinkage-reducing admixture in reducing autogenous shrinkage stress of ultra-high-performance fiber-reinforced concrete. *Cem. Concr. Compos.* **2015**, *64*, 27–36. [CrossRef]
15. Shen, D.; Jiang, J.; Shen, J.; Yao, P.; Jiang, G. Influence of curing temperature on autogenous shrinkage and cracking resistance of high-performance concrete at an early age. *Constr. Build. Mater.* **2016**, *103*, 67–76. [CrossRef]
16. Negahban, E.; Bagheri, A.; Sanjayan, J. One-Year study of restrained shrinkage and creep behaviours of geopolymer concrete. *Constr. Build. Mater.* **2023**, *376*, 131057. [CrossRef]
17. Khan, I.; Xu, T.; Castel, A.; Gilbert, R.I. Early-age tensile creep and shrinkage-induced cracking in internally restrained concrete members. *Mag. Concr. Res.* **2019**, *71*, 1167–1179. [CrossRef]
18. Gao, Z. Research on Evaluation Technique of Concrete Cracking Performance under Restraint Condition. *For. Chem. Rev.* **2022**, 1034–1041. Available online: https://www.forestchemicalsreview.com/index.php/JFCR/article/view/986 (accessed on 1 August 2023).
19. Huang, L.; Hua, J.; Kang, M.; Luo, Q.; Zhou, F. Influence of Steel Plates and Studs on Shrinkage Behavior and Cracking Potential of High-Performance Concrete. *Materials* **2019**, *12*, 342. [CrossRef]
20. Chen, L.; Chen, Z.; Zhang, Y.; Liu, Y.; Osman, A.; Farghali, M.; Hua, J.; Al-Fatesh, A.; Ihara, I.; Rooney, D.; et al. Artificial intelligence-based solutions for climate change: A review. *Environ. Chem. Lett.* **2023**. [CrossRef]
21. Chen, L.; Huang, L.; Hua, J.; Chen, Z.; Wei, L.; Osman, A.; Fawzy, S.; Rooney, D.; Dong, L.; Yap, P. Green construction for low-carbon cities: A review. *Environ. Chem. Lett.* **2023**, *21*, 1627–1657. [CrossRef]
22. Chen, L.; Msigwa, G.; Yang, M.; Osman, A.; Fawzy, S.; Rooney, D.; Yap, P. Strategies to achieve a carbon neutral society: A review. *Environ. Chem. Lett.* **2022**, *20*, 2277–2310. [CrossRef] [PubMed]
23. *JGJT380-2015*; Technical Specification for Steel Plate Shear Walls. Ministry of Housing and Urban-Rural Development of the People's Republic of China: Beijing, China, 2016.
24. *GB50010-2010*; Concrete Structural Design Code. Ministry of Housing and Urban-Rural Development of the People's Republic of China: Beijing, China, 2015.

25. *JGJ3-2010*; Technical Specification for High-Rise Buildings with Concrete Structures. Ministry of Housing and Urban-Rural Development of the People's Republic of China: Beijing, China, 2010.
26. *JGJ138-2016*; Combined Structural Design Specification. Ministry of Housing and Urban-Rural Development of the People's Republic of China: Beijing, China, 2016.
27. *GB50011-2010*; Seismic Design Code for Buildings. Ministry of Housing and Urban-Rural Development of the People's Republic of China and General Administration of Quality Supervision, Inspection and Quarantine of the People's Republic of China: Beijing, China, 2016.
28. *GBT50081-2019*; Test Methods of Physical and Mechanical Properties of Concrete. Ministry of Housing and Urban-Rural Development of the People's Republic of China and the State Administration of Market Supervision and Administration of the People's Republic of China: Beijing, China, 2019.
29. *GB50017-2017*; Steel Structure Design Standard. Ministry of Housing and Urban-Rural Development of the People's Republic of China: Beijing, China, 2017.

buildings

MDPI

Article

Moisture Diffusion Coefficient of Concrete under Different Conditions

Fengbin Zhou [1], Wenhao Li [1], Ying Hu [1], Lepeng Huang [1,*], Zhuolin Xie [1], Jun Yang [2], Daifeng Wu [2] and Zhonghao Chen [3]

1 School of Civil Engineering, Chongqing University, Chongqing 400045, China; 20141613158@cqu.edu.cn (F.Z.); wenhaoli@cqu.edu.cn (W.L.); huying@stu.cqu.edu.cn (Y.H.); zhuolin.xie@cqu.edu.cn (Z.X.)
2 Chongqing Urban Investment Infrastructure Construction Co., Ltd., Chongqing 400045, China; 15215048656@163.com (J.Y.); 13896090632@139.com (D.W.)
3 Department of Civil Engineering, Xi'an Jiaotong-Liverpool University, Suzhou 215123, China; zhonghao.chen21@alumni.xjtlu.edu.cn
* Correspondence: huang_lepeng@cqu.edu.cn

Abstract: Humidity change in concrete is the leading cause of concrete shrinkage. Moreover, the moisture diffusion coefficient of concrete is an essential parameter for assessing and predicting the internal moisture content of concrete. However, there is a lack of theoretical construction and experimental studies on the effect of different conditions, especially different constraints, on the moisture diffusion coefficient of concrete. Therefore, the internal humidity, pore structure parameters, and basic mechanical properties of concrete under different strength grades C30, C40, C50, and C60 (C stands for concrete and numbers indicate the strength class of the concrete), curing environments (dry and sealed curing conditions), and constraints were tested in this study. In addition, a calculation model of concrete's internal humidity and humidity diffusion coefficient was established. The research findings show that the internal humidity of concrete decreased with age due to hydration and drying. External humidity had a significant effect on the moisture change of concrete, and the lower the external humidity, the larger the humidity diffusion coefficient and the faster the internal humidity of concrete decreases. Reinforcement (confinement) changes the pore structure parameters of the concrete, which in turn affects the transport of moisture within the concrete. The higher the reinforcement rate, the larger the pore structure parameters of the concrete, the larger the humidity diffusion coefficient, and the faster the concrete humidity decreases. The method proposed in the study can accurately predict the internal humidity of concrete using the humidity diffusion coefficient. The research results are a reference for preventing concrete shrinkage and cracking in construction.

Keywords: internal concrete moisture; concrete moisture diffusion coefficient; pore structure

1. Introduction

Concrete has become the most widely utilized construction material worldwide due to its advantages, namely ease of access to materials, ease of molding, durability, low energy consumption, and ease of compositing with other materials. The inspection results of the US National Cooperative Highway Research Program indicate that in 1995, approximately 100,000 concrete bridge deck slabs exhibited penetrating cracks at 1–3 m intervals within one month after the concrete was poured [1]. In China, the problem of concrete cracking is crucial. Survey results indicate that for transportation facilities, ranging from bridges and pavements to seaport engineering structures along the coast and from civil and commercial buildings to industrial buildings, concrete cracking is exceedingly prevalent, and this research problem has attracted extensive attention from the engineering community in recent years [2–4]. China's concrete structures have been exposed to numerous problems associated with cracking during construction and operation, and the service life of many concrete structures is much lower than their design life, with cracks appearing within 3–5 years in some scenarios, and in other scenarios, the cracks exceeded the permissible crack widths even before the concrete structures were formally utilized, causing immense economic losses [5,6].

The shrinkage deformation of concrete Is one of the main characteristics of early-age concrete, and the low water-to-cement ratio, high compactness, high fluidity, and large amount of mineral admixtures exhibited by modern concrete may further increase the early deformation of concrete [7–10]. Due to the continuous improvements in the mechanical properties of concrete materials and structural design methods, cracks caused by direct loading rarely occur in concrete structures under normal usage [11–13]. The mainstream structural design method is largely based on the strength criterion for the design of structures, and concrete is regarded as a uniform, stable material with fixed strength. Therefore, its early instability and deformation characteristics are not considered in the design; consequently, the indirect effects (i.e., shrinkage, temperature, and deformation) considerably influence concrete cracking.

According to capillary pore stress theory, concrete shrinkage is mainly attributable to the capillary pore stress produced on the pore wall after the internal water is consumed [14]. Due to cement hydration and humidity diffusion, the internal humidity of concrete gradually decreases. Once the relative humidity begins to decrease, capillary pore tension is triggered, and the scale of the capillary pore tension then controls the macroscopic shrinkage deformation of the concrete; thus, the internal humidity change rule of concrete is necessary for analyzing concrete shrinkage deformation [15,16]. In an experimental study on the internal humidity of concrete, Persson [17] measured the changes in the humidity of concrete specimens exposed to air, immersion curing, and sealed conditions and established a simulation formula based on the water–cement ratio, cement hydration, or age. Kim et al. [18] used the sealed dry specimens to measure the humidity diffusion of concrete with different water–cement ratios and the internal humidity changes caused by cement hydration. The results indicated that the humidity decrease in concrete with a high water–cement ratio was mainly due to humidity diffusion, and the humidity decrease caused by cement hydration in concrete with a low water–cement ratio was more significant. Nillsson et al. [19] tested and measured the change rule associated with the relative humidity of concrete in seawater and natural environments. Andrade et al. [20] tested concrete exposed to outdoor temperature and humidity conditions. Parrot [21] and Nillsso et al. [19] experimentally measured the relative humidity in concrete specimens exposed to the natural environment or seawater.

The moisture diffusion coefficient is commonly utilized by researchers to predict the variation pattern of moisture inside concrete [22,23]. Most researchers have utilized a linear diffusion equation to predict moisture diffusion [24–26], which considers the moisture diffusion coefficient as a constant, and the obtained results exhibit some deviation from the experimental tests [27]. The rate of pore water migration in concrete is dependent on the pore structure characteristics and the cement paste distribution. Moreover, it is also closely related to the water in the pore space [28–30]; therefore, the moisture diffusion in concrete is highly nonlinear. Experimental studies have confirmed that the moisture diffusion coefficient is a function of the water–cement ratio and the maturity of the concrete and that linear equations are not applicable to the solution of the moisture transport problem of concrete [31]. Ayano [32] utilized the slice weighing method to obtain the water content of specimens at different ages and positions and assumed that the diffusion coefficient was an exponential equation related to the loss of water dispersion; thus, the researcher solved the nonlinear diffusion equation by combining the residual method and the nonlinear least squares method. Sakata and Kuramoto [33] and Akita et al. [34] calculated the diffusion coefficient by performing the Boltzmann transform on the diffusion equation, and the results were consistent with the experimental results obtained by Bazant and Najjar [31] using the slice method. In addition to the aforementioned studies, some scholars have recently utilized nonlinear diffusion equations to predict the moisture diffusion pattern of concrete [35–37]. However, the number of studies is overall insufficient for identifying a method with generalization to the vast majority of cases. Therefore, research on the theoretical solution of the concrete moisture field using a nonlinear method to validate the accuracy of the method under as many environmental conditions as possible is required.

Notably, the effect of confinement on moisture diffusion in concrete has been neglected by researchers. In engineering practice, concrete is always restrained by reinforcement. Studies have revealed that when lateral restraints are used to limit the expansion of concrete containing an expansive agent, the porosity, average pore diameter, and porosity content of concrete with lateral restraint can be reduced compared to those without lateral restraint [38,39]. Reinforcing steel inhibits the shrinkage effect of concrete and reduces the volume reduction due to shrinkage [40–45]. Therefore, a reasonable inference is that reinforcement changes the pore structure of concrete under shrinkage. However, at present, the exact effect is not clear. The pore structure of concrete is an important channel for moisture transport, and changes in the pore structure will likely lead to changes in the moisture transport pattern of concrete. Therefore, research on the effect of steel reinforcement on the pore structure of concrete and the moisture transport law is crucial.

Therefore, in this study, the internal humidity, pore structure, and basic mechanical properties of concrete with different strength grades (C30, C40, C50, and C60), curing environments (dry curing and sealing curing), and constraints were tested using a self-developed testing device, and the calculation model for the internal humidity of concrete and the moisture diffusion coefficient was established. The study revealed the changing rule of the humidity diffusion coefficient of concrete under different external environmental conditions and constraints and elucidated the effect of the internal pore structure of concrete on the moisture transport characteristics of concrete.

2. Theoretical Modeling

The method of calculating the moisture diffusion coefficient is derived in this study to analyze the moisture of concrete. Due to the rate of cement hydration and water diffusion in concrete being relatively slow, the relationship between internal humidity and water content in concrete can be approximated as linear [31,34,46,47]. Under test conditions, concrete undergoes humidity exchange in three dimensions (x, y, and z); thus, the humidity transport is three-dimensional. To simplify the calculation, the concrete is regarded as homogeneous and the humidity transport to be the same in the x, y, and z directions. Therefore, the moisture diffusion equation inside concrete due to drying action can be written as follows based on Fick's second law:

$$\frac{\partial H_d}{\partial t} = \frac{\partial}{\partial x}(3D\frac{\partial H_d}{\partial x}) = 3D\frac{\partial^2 H_d}{\partial x^2} \tag{1}$$

where H_d denotes the humidity reduction due to the drying effect, t denotes the drying time, x denotes the distance of the interior of the concrete from the drying surface, and D denotes the concrete diffusion coefficient.

In the past, Equation (1) has been solved using the finite difference method, which simplifies the solution process; however, the interpretation of its calculation results is more complicated and should be facilitated by corresponding software programming. Therefore, this study attempts to solve Equation (1) using a Fourier transform.

Herein, only the outward diffusion of the concrete humidity is considered: the reverse of the humidity into the concrete is ignored. Therefore, in the humidity diffusion process, which follows the inside concrete–outside concrete direction, the concrete diffusion coefficient D must be greater than 0. Subsequently, let

$$3D = a^2 \tag{2}$$

where a denotes the parameter used for the calculation.

The H_d can be expressed as:

$$H_d = u + H_e \tag{3}$$

where u denotes the change in humidity inside the concrete due to moisture diffusion, and H_e denotes the ambient humidity of the concrete.

The concrete internal humidity is a function of space and time H_d (x,t); therefore, its boundary conditions can be expressed as follows:

$$\begin{cases} H_d(x,0) = H_0 \\ H_d(0,t) = H_e \\ H_d(\infty,t) = H_0 \end{cases} \tag{4}$$

where H_0 denotes the initial humidity inside the coagulation.

Subsequently, the boundary condition of Equation (3) is expressed as follows:

$$\begin{cases} u(x,0) = H_0 - H_e \\ u(0,t) = 0 \\ u(\infty,t) = H_0 - H_e \end{cases} \tag{5}$$

Subsequently, the following equation is obtained:

$$\begin{cases} u_t - a^2 u_{xx} = 0 & (0 < x < \infty, t > 0) \\ u(x,0) = H_0 - H_e & (0 \le x < \infty) \\ u(0,t) = 0 \end{cases} \tag{6}$$

where u_t denotes the first-order partial derivative of the internal concrete humidity with respect to t and u_{xx} denotes the second-order partial derivative of the internal concrete humidity with respect to x.

Notably, Equation (6) represents a semi-infinite boundary problem, and to solve it, it should be transformed into an infinite boundary problem, which can be expressed as follows:

$$U = \begin{cases} u & 0 < x < \infty \\ u & -\infty < x < 0 \end{cases} \tag{7}$$

where U denotes the change in humidity due to humidity diffusion considering infinite boundary conditions.

Subsequently, we obtain

$$U(x,0) = \varphi(x) = \begin{cases} H_0 - H_e & 0 < x < \infty \\ 0 & x = 0 \\ -H_0 + H_e & -\infty < x < 0 \end{cases} \tag{8}$$

where $\varphi(x)$ denotes the change in humidity at each location at the initial moment due to humidity diffusion.

Consequently, the following equation is obtained:

$$\begin{cases} U_t - a^2 U_{xx} = 0 & (-\infty < x < \infty, t > 0) \\ U(x,0) = \varphi(x) & (-\infty \le x < \infty) \end{cases} \tag{9}$$

Setting

$$U(x,t) = X(x)T(t) \tag{10}$$

where U_t denotes the first-order partial derivative of the internal moisture diffusion of concrete with respect to t under infinite boundary conditions, U_{xx} denotes the second-order partial derivative of the internal moisture diffusion of concrete with respect to x under infinite boundary conditions, $X(x)$ denotes the change in humidity induced by the internal moisture diffusion of concrete on the x scale under infinite boundary conditions, and $T(t)$ denotes the change in humidity induced by the internal moisture diffusion of concrete on the t scale under infinite boundary conditions.

Subsequently,

$$X(x)\frac{\partial T}{\partial t} - a^2\frac{\partial^2 X}{\partial x}T(t) = 0 \tag{11}$$

Namely,

$$\frac{T\prime}{T} = a^2\frac{X''}{X} \tag{12}$$

Setting

$$a^2\frac{X''}{X} = a^2\kappa \tag{13}$$

where κ denotes the computational parameter.

Consequently,

$$\begin{cases} T' + \kappa a^2 T = 0 \\ X'' + \kappa X = 0 \end{cases} \tag{14}$$

From Equation (14), we obtain the following equation:

$$T = Ae^{-\kappa a^2 t} \tag{15}$$

Order

$$\kappa = \alpha^2 \tag{16}$$

Subsequently, the solution of Equation (14) is as follows:

$$X = A_1 \cos \alpha x + B_1 \sin \alpha x \tag{17}$$

At this point, α can assume any value, and U_α satisfies the superposition principle, namely

$$U = \sum U_\alpha \tag{18}$$

Therefore,

$$U_\alpha = \int_{-\infty}^{\infty} U_\alpha d\alpha = \int_{-\infty}^{\infty} (A_1 \cos \alpha x + B_1 \sin \alpha x)e^{-\alpha^2 a^2 T} d\alpha \tag{19}$$

where U_α denotes the change in humidity induced by the diffusion of humidity within the concrete on the α scale under infinite boundary conditions.

From the initial conditions, when $t = 0$,

$$\int_{-\infty}^{\infty} U_\alpha d\alpha = \int_{-\infty}^{\infty} (A_1 \cos \alpha x + B_1 \sin \alpha x)e^{-\alpha^2 a^2 T} d\alpha = \varphi(x) \tag{20}$$

From Fourier transforms, we obtain the following equations:

$$A_1 = \frac{1}{2\pi}\int_{-\infty}^{\infty} \cos \alpha x \varphi(x) d\alpha \tag{21}$$

$$B_1 = \frac{1}{2\pi}\int_{-\infty}^{\infty} \sin \alpha x \varphi(x) d\alpha \tag{22}$$

where $A1$, $B1$ denote the Fourier coefficients of the function U_α.

Subsequently,

$$\begin{aligned} U(x,t) &= \frac{1}{2\pi}\int_{-\infty}^{\infty}\varphi(\zeta)[\int_{-\infty}^{\infty}(A_1\cos\alpha x + B_1\sin\alpha x)e^{-\alpha^2 a^2 T}d\alpha]d\zeta \\ &= \frac{1}{2\pi}\int_{-\infty}^{\infty}\varphi(\zeta)e^{-\frac{(\zeta-x)^2}{4at}}d\zeta \end{aligned} \quad (23)$$

Further simplification yields the following expression:

$$U(x,t) = \frac{2(H_0 - H_e)}{\sqrt{\pi}}\int_0^{\frac{x}{2a\sqrt{t}}} e^{-\zeta^2}d\zeta \quad (24)$$

The above Equation (24) includes the term "error function" which is a nonfundamental function in mathematics, namely

$$erf(x) = \frac{2}{\sqrt{\pi}}\int_0^x e^{-\zeta^2}d\zeta \quad (25)$$

Subsequently, Equation (24) can be expressed as

$$U(x,t) = (H_0 - H_e)erf(\frac{x}{2a\sqrt{t}}) \quad (26)$$

Consequently, we obtain

$$u = U(0 < x < \infty) = (H_0 - H_e)erf(\frac{x}{2a\sqrt{t}}) \quad (27)$$

From the assumptions of Equation (3), the following equation is obtained:

$$H_d = H_e - (H_0 - H_e)erf(\frac{x}{2\sqrt{3Dt}}) \quad (28)$$

Considering the form of the error function, the error function does not exhibit an original function and cannot be solved by the classical Newton–Leibniz formula, but only an approximate calculation can be performed. Currently, commonly utilized computational methods include the complex trapezoidal product method [10,48] and the hyperbolic tangent function approximation method [49]. Herein, the results obtained from the hyperbolic tangent function approximation calculation method are utilized to approximate the error function; subsequently, Equation (28) can be expressed as follows:

$$H_d = H_e - (H_e - H_0)\tanh[1.12838\frac{x}{2\sqrt{3Dt}} + 0.10277(\frac{x}{2\sqrt{3Dt}})^3] \quad (29)$$

Therefore, if the initial internal humidity, external ambient humidity, and the humidity diffusion coefficient of the concrete are obtained, the humidity distribution inside the concrete can be calculated using Equation (29).

Generally, the ambient humidity to which the concrete is subjected and the initial internal humidity are easily obtained. Therefore, to solve Equation (29), the value of the moisture diffusion coefficient D for concrete should be determined.

Synthesizing existing research results, the Boltzmann transform is applied herein to solve the diffusion coefficient inversely. When the concrete is dry on both sides except in the x-direction (no water loss on the bottom surface of the specimen), based on the linear relationship between internal humidity and moisture in concrete and Fick's second law, the following expression is applicable:

$$\frac{\partial H_d}{\partial t} = \frac{\partial}{\partial x}(D\frac{\partial H_d}{\partial x}) + 2\frac{\partial}{\partial y}(D\frac{\partial H_d}{\partial y}) + 2\frac{\partial}{\partial z}(D\frac{\partial H_d}{\partial z}) \tag{30}$$

Introducing the variable η

$$\eta = \frac{x}{\sqrt{t}} \tag{31}$$

Let $y = \alpha x$ and $z = \beta x$; subsequently, Equation (30) can be expressed as follows:

$$-\frac{1}{2}\eta = (1 + 2\alpha^{-2} + 2\beta^{-2})\frac{\partial}{\partial H_d}[D(H_d)\frac{\partial H_d}{\partial \eta}] \tag{32}$$

Integrating Equation (32) between H and H_0 yields the following equation:

$$D = [\int_{H}^{H_0} \frac{1}{2}\eta dH_d]\left[(1 + 2\alpha^{-2} + 2\beta^{-2})\frac{\partial H_d}{\partial \eta}|_H\right]^{-1} \tag{33}$$

From the test conditions, the dimensions of the specimen in the x, y, and z directions are 200 mm × 200 mm × 1000 mm; subsequently, we obtain $\alpha = 1$ and $\beta = 5$.

Consequently, the moisture diffusion coefficient of concrete under test conditions can be expressed as follows:

$$D = \frac{1}{3.08}\int_{H}^{H_0} \frac{1}{2}\eta dH_d\left[\frac{\partial H_d}{\partial \eta}|_H\right]^{-1} \tag{34}$$

Therefore, from the experimentally obtained H_d and the relationship between H_d and η, the humidity diffusion coefficient of the concrete can be calculated using Equation (34), as can the internal humidity of the concrete.

In previous studies, the variation in internal humidity with age under sealed and dry conditions was experimentally obtained for each concrete grade. Therefore, the moisture diffusion coefficient of concrete can be obtained by inverting Equation (34). The decrease in humidity in concrete is due to a combination of cement hydration and drying effects. Therefore, in calculating the moisture diffusion coefficient of concrete, the moisture drop due to concrete hydration should be deducted from the total moisture reduction.

Because the discrete humidity values obtained from the tests cannot be directly substituted into Equation (34) to invert the moisture diffusion coefficient of concrete, this study proposes a mathematical model for the relationship between H_d and η based on Akita's research results [34]; by approximating the fitting of the test results, the humidity value is calculated as follows:

$$H_d = H_0\left[1 + a - \frac{b}{(0.5\eta + c)^d}\right] \tag{35}$$

where a, b, c, and d denote the coefficients of the model. These coefficients can be obtained by fitting the test results.

To validate theoretical model, we conduct experiments in the next section.

3. Experimental Design

3.1. Concrete Mixing Ratio

Past studies have confirmed that the various aspects of the performance of concrete of different strengths differ significantly. Therefore, four commonly used concrete strengths (C30, C40, C50, and C60) were selected to study the variation in their internal humidity and humidity diffusion coefficient under different conditions. Cement, water, fly ash, sand, coarse aggregate, and water reducing agent were used to prepare the concrete used for the experiment and the amount of these materials required per cubic meter of the mixture was prepared, as shown in Table 1. The percentage content of SiO_2, CaO, Al_2O_3, Fe_2O_3, MgO, and SO_3 in cement and fly ash is shown in Table 2. The specific surface area of cement and fly ash were 3471 and 4680 cm^2/g. The density was 3.10 and 2.22 g/cm^3, respectively.

Table 1. Concrete mix proportion.

Concrete	Cement (kg/m^3)	Water (kg/m^3)	Fly Ash (kg/m^3)	Sand (kg/m^3)	Coarse Aggregate (kg/m^3)	Water Reducing Agent (kg/m^3)
C30	336	195	69	705	1094	9.9
C40	365	185	65	685	1090	10.4
C50	394	175	61	665	1086	10.9
C60	423	166	58	645	1083	11.4

Table 2. Chemical compositions and physical properties of the cementitious materials.

Composition (%)	Cement	Fly Ash
SiO_2	21.47	49.47
CaO	65.77	4.45
Al_2O_3	5.47	20.67
Fe_2O_3	4.28	14.32
MgO	1.44	1.17
SO_3	0.52	1.40

3.2. *Basic Concrete Performance Tests*

To understand the basic mechanical properties of the ordinary and high-strength concrete utilized herein, this study was conducted according to GB/T 50081-2002 [50]; thus, the cubic compressive strength, split tensile strength, and static compressive elastic modulus of the concrete were tested.

The basic mechanical properties of concrete specimens were tested after 3, 7, and 28 days of curing. The specimens used for the basic mechanical tests were subjected to the same curing conditions as those used for the moisture tests.

The test results of cubic compressive strength, split tensile strength, and static modulus of elasticity of the concrete samples with four mixing ratios at different ages are listed in Table 3.

Table 3. Basic mechanical properties of concrete.

Basic Properties	Concrete Strength	Curing Condition	Age (Days)		
			3	7	28
Cube compressive strength (MPa)	C30	Dry	15.3	21.3	34.5
		Sealed	17.1	21.4	39.5
	C40	Dry	25.1	31.0	45.4
		Sealed	27.0	32.0	48.5
	C50	Dry	34.0	40.6	55.0
		Sealed	36.5	42.0	56.5
	C60	Dry	44.0	53.0	63.0
		Sealed	46.5	59.4	68.0
Splitting tensile strength (MPa)	C30	Dry	2.2	3.7	4.3
		Sealed	2.4	4.1	4.4
	C40	Dry	2.5	4.1	4.7
		Sealed	2.7	4.2	4.7
	C50	Dry	2.9	4.4	5.0
		Sealed	3.1	4.5	5.1
	C60	Dry	3.4	4.7	5.4
		Sealed	3.5	4.9	5.6

Table 3. *Cont.*

Basic Properties	Concrete Strength	Curing Condition	Age (Days)		
			3	7	28
Static modulus of elasticity ($\times 10^4$ MPa)	C30	Dry	1.6	1.7	2.9
		Sealed	1.7	2.0	2.9
	C40	Dry	1.8	2.1	3.0
		Sealed	1.6	2.2	3.0
	C50	Dry	2.0	2.3	3.1
		Sealed	2.1	2.4	3.2
	C60	Dry	2.5	2.5	3.2
		Sealed	2.4	2.6	3.4

3.3. Concrete Internal Humidity and Shrinkage

The molds used to form the specimens discussed in this chapter were composed of Plexiglas. The specimen dimensions were 200 mm × 200 mm × 1000 mm (width, height, and length). Meanwhile, two pieces of 1 mm-thick Teflon film were placed at the bottom of the mold to reduce the friction between the bottom of the specimen and the concrete.

The humidity inside the concrete was measured using a humidity sensor with a 3% accuracy for relative humidity. The humidity sensor is a moisture-sensitive capacitance type, which comprises polymer film capacitors, such as polystyrene, polyimide, and caseinate acetate fiber. When the ambient humidity is altered, the dielectric constant of the moisture-sensitive capacitor changes; thus, its capacitance also changes in proportion to the relative humidity, from which the change in external humidity can be obtained. The humidity sensors were placed 100 mm from the top surface of the concrete to test the humidity changes inside the concrete. To install the humidity sensor inside the concrete, a PVC pipe with an inner diameter of 20 mm and a thin PVC sheet covering the bottom end were used. Two rectangular holes were cut close to the bottom of the PVC pipe to allow the sensors to test the humidity inside the concrete. To prevent fresh concrete from flowing into the PVC pipe during the pouring process, a thin steel pipe with an outer diameter of 20 mm was inserted into the PVC pipe before pouring the concrete. The PVC and the thin steel pipes were inserted 100 mm from the top surface of the mold before the concrete was poured. When the concrete started to set, the thin steel pipe was pulled out, and the humidity sensor was inserted into the PVC pipe. To prevent humidity exchange between the inside of the PVC pipe and the outside environment, the top of the PVC pipe was sealed with a sealant and a polyethylene film (Figure 1).

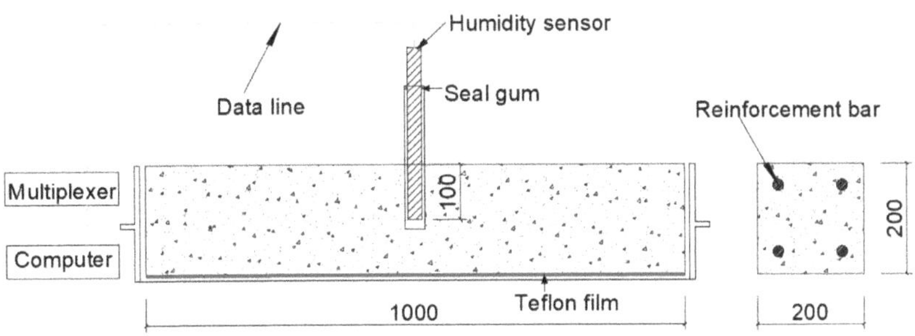

Figure 1. RH test setup for the concrete (all the dimensions are in mm).

To measure the internal humidity and determine pore structure change rule of concrete under different curing and external environments, two types of concrete curing methods were designed, namely dry curing and sealed curing. For dry components, the external humidity was set to 35, 50, and 65%. For the dry curing specimens, after the concrete was first set and demolded and the installation of humidity sensors, the test began. For the sealed curing specimens, immediately after the installation of the moisture sensor, the whole specimen was sealed on six sides with aluminum foil tape to prevent the exchange of moisture between the concrete and the outside environment; thus, the humidity change due to cement hydration was obtained.

To investigate the effect of constraints on the moisture of concrete, unconfined and confined specimens were used in this study. In the restrained specimens, four reinforcement bars were placed into four sides of the cross-sections of the restrained specimen to meet the reinforcement ratio requirements (Figure 1). The modulus of elasticity of the steel bar was 20.6×10^4 MPa, the yield strength was 400×10^4 MPa, and the ultimate strength was 580×10^4 MPa. Specific experimental information is tabulated in Table 4 (without constraints) and Table 5 (with constraints). All the components were tested at a room temperature of 25 °C. The internal humidity of the concrete test lasted for a total of 28 days. The internal humidity data of the concrete were automatically read and recorded by a computer every 0.5 days. Three specimens of each type were evaluated, and the data analyzed in the following sections were derived from the average of the data obtained from three specimens under the same test conditions.

Table 4. List of experiments under unconstrained conditions.

No.	Concrete Strength	External Humidity (%)	Curing Condition
C30-35	C30	35	Dry
C30-50	C30	50	Dry
C30-65	C30	65	Dry
C30-Sealed	C30	-	Sealed
C40-35	C40	35	Dry
C40-50	C40	50	Dry
C40-65	C40	65	Dry
C40-Sealed	C40	-	Sealed
C50-35	C50	35	Dry
C50-50	C50	50	Dry
C50-65	C50	65	Dry
C50-Sealed	C50	-	Sealed
C60-35	C60	35	Dry
C60-50	C60	50	Dry
C60-65	C60	65	Dry
C60-Sealed	C60	-	Sealed

Table 5. List of experiments under constraints.

No.	Concrete Strength	External Humidity (%)	Reinforcement Ratio (%)	Curing Condition	Reinforcement Bar Diameter (mm)
C30-65-1	C30	65	1.14	Dry	12
C30-65-3	C30	65	3.24	Dry	20
C30-65-6	C30	65	6.56	Dry	28
C60-65-1	C60	65	1.14	Dry	12
C60-65-3	C60	65	3.24	Dry	20
C60-65-6	C60	65	6.56	Dry	28

3.4. Pore Structure Test

This study was conducted to test the pore structure parameters of concrete of different strength classes and different ages (3 days, 7 days, and 28 days) to reveal the relationship between pore structure and humidity and humidity diffusion coefficient. The pore structure of concrete specimen blocks can be analyzed using mercury-in-pressure (MIP); thus, some concrete pore structure parameters, such as porosity, mean pore size, and median pore size can be obtained.

For porosity testing, the samples used were obtained from specimens of the same concrete grade and curing conditions as the specimens tested for internal moisture. When the test age was attained, the specimen was removed, and the concrete on the specimen was drilled with a drilling machine. Afterward, the specimen was crushed, and after removing the coarse aggregate, it was placed in acetone to terminate the cement hydration process. Before starting the test, the specimens were dried in a high-temperature vacuum (>150 °C) and placed in a desiccator. The test results are derived from the average of the test results of the six test samples.

4. Results and Discussion

4.1. Concrete Internal Humidity

Figure 2 demonstrates the change rule of internal humidity with age for all specimens. Overall, the change in the internal humidity of concrete can be divided into two stages: Stage I (rapid decline in humidity) mainly occurs in concrete after pouring for 0–8 days. In this stage, the humidity in the concrete continued to decline rapidly. Stage II (slowly decreasing humidity) mainly occurred 8–28 days after concrete casting, in which the rate of decreasing humidity of concrete decreased and eventually stabilized. This phenomenon can be explained as follows: In the early age of concrete, the water in concrete was rapidly consumed due to intense hydration. Meanwhile, the drying effect was intense due to the large gradient of humidity between the interior and exterior of the concrete. Under the two effects, the internal humidity of concrete decreases rapidly. With the gradual slowing down of hydration and decrease in the internal and external humidity gradient, the water consumption rate in the concrete slowed down, and the concrete humidity decreased at a lower rate and gradually stabilized. Regarding sealed elements, the elements were sealed with aluminum foil tape. The water inside the component was not exchanged with the outside environment, and only the hydration of the cement inside the concrete consumed the water in the concrete. The hydration of high-strength-grade concrete was more intense due to its lower water–cement ratio. As a result, the rate of decrease of humidity inside the concrete increased with the increase in cement strength, and at the age of 28 days, the humidities of C30, C40, C50, and C60 sealed members were 93, 92, 87, and 85%, respectively.

In drying members, in addition to the consumption of water by the hydration of cement, the drying action also consumes the water inside the concrete, which makes the humidity of the concrete decrease faster and the final humidity lower. From the experimental results, the lower the humidity of the external environment, the stronger the drying effect and the faster the humidity of concrete decreases. Taking C30 concrete as an example, when the external humidity was 35, 50, and 65%, the internal humidity of the concrete specimen was 56, 70, and 74%, respectively, at 28 days.

4.2. Concrete Moisture Diffusion Coefficient

The concrete moisture diffusion coefficients for specimens C30-35, C30-50, C30-65, C40-35, C40-65, C50-35, C50-50, C60-35, C60-50, C60-65, C30-65-1, C30-65-3, C30-65-6, C60-65-1, C60-65-3, and C60-65-6 can be calculated using Equations (34) and (35), where H_0, a, b, c, and d represent calculation parameters of the humidity value fitting model (Equation (35)), and it can be obtained from the results of the fitting tests, as shown in Table 6. In addition, average fitting error is obtained between the experimental and fitted values of humidity diffusion at each age of the specimens.

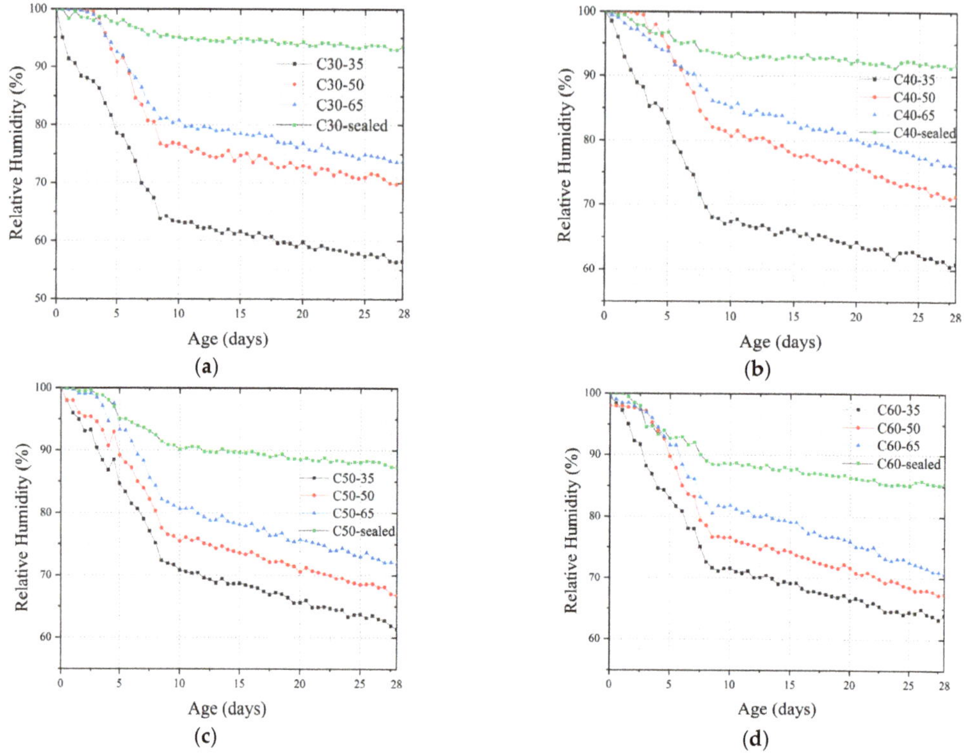

Figure 2. Changes in the internal humidity of concrete: (**a**) C30; (**b**) C40; (**c**) C50; (**d**) C60.

Table 6. Calculation model parameters.

No.	H_0	a	b	c	d	Average Fitting Error
C30-35	98.7	0.020	8.1	3.8	1.7	1.1%
C30-50	98.5	0.014	7.6	3.5	1.9	0.3%
C30-65	99.0	0.015	16	4.5	2.3	2.1%
C40-35	98.5	0.015	8.5	3.0	2.0	0.9%
C40-50	98.5	0.010	16.0	3.6	2.4	4.6%
C40-65	98.6	0.013	21.0	3.5	2.6	6.5%
C50-35	99.1	0.005	40.0	4.0	2.8	1.4%
C50-50	98.5	0.008	18.0	3.8	2.6	0.7%
C50-65	98.6	0.009	20.0	3.5	2.7	4.1%
C60-35	98.8	0.010	15.0	2.5	2.6	2.9%
C60-50	99.4	0.012	15.0	3.0	2.6	3.1%
C60-65	99.6	0.015	15.0	3.2	2.6	5.2%
C30-65-1	98.5	0.011	15.0	6.0	2.3	6.2%
C30-65-3	99.4	0.010	19.0	6.0	2.3	4.2%
C30-65-6	99.5	0.011	25.0	6.0	2.3	2.2%
C60-65-1	98.5	0.016	10.0	3.0	2.3	0.05%
C60-65-3	99.0	0.016	12.0	4.0	2.3	2.9%
C60-65-6	99.0	0.015	13.0	5.0	2.3	5.6%

Figure 3 shows the variation rule of the moisture diffusion coefficient of concrete with age under different external environments and strength grades. The test results reveal that the moisture diffusion coefficient of concrete changes with age. Its age-based change rule can be divided into three stages. Stage I (rapid decline stage of diffusion coefficient):

After going through the stage of 100% humidity in the previous period, the humidity of the concrete starts to decrease rapidly, which is from 0 days to 1 day after the concrete cures, during which the humidity diffusion coefficient of the concrete rapidly decreases. Stage II (slow decline stage): this period occurs from 1 day to 8 days after concrete placement. Stage III (stabilization stage): when the age of concrete exceeds 8 days, the change of concrete humidity diffusion coefficient is no longer significant for both ordinary and high-strength concrete.

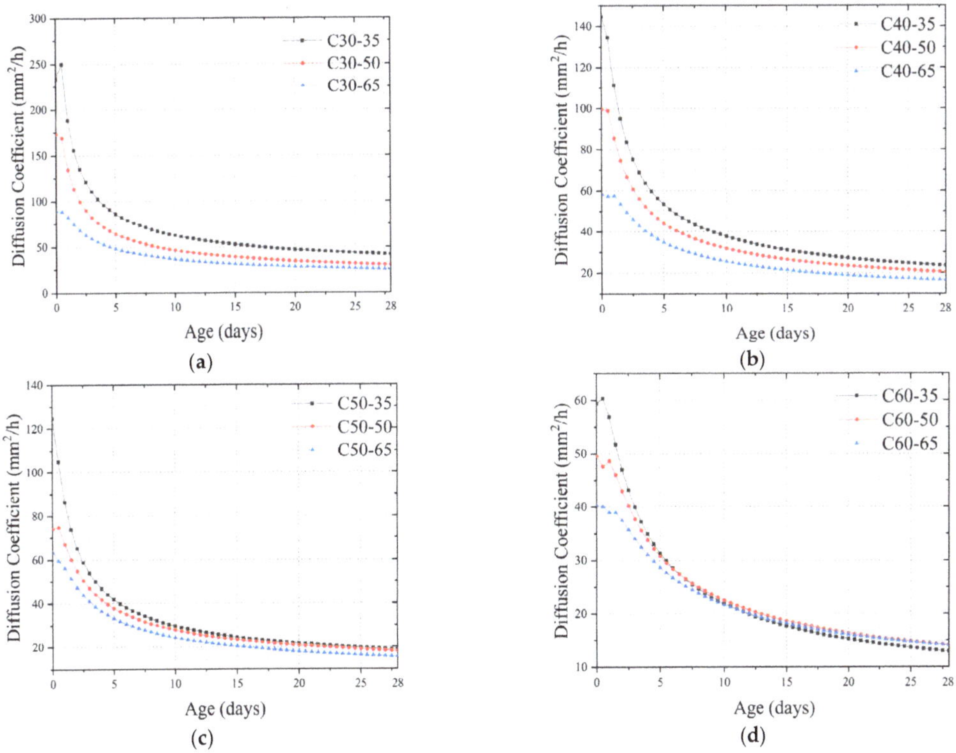

Figure 3. Development of the moisture diffusion coefficient of concrete with age: (**a**) C30; (**b**) C40; (**c**) C50; (**d**) C60.

The external humidity environment has a significant effect on the moisture diffusion coefficient of concrete. When the external humidity of concrete is lower, the moisture diffusion coefficient of concrete is higher. Taking C30 concrete as an example, at the age of 3 days, the humidity diffusion coefficient of concrete in a 35% external humidity environment is 110.5 mm^2/h. With the increase in external humidity to 50% and 65%, the humidity diffusion coefficient decreases to 68.9 mm^2/h and 53.8 mm^2/h, which is a decrease of 41.6 mm^2/h and 56.7 mm^2/h, respectively. This phenomenon occurs because when the concrete is poured, the internal humidity is significantly higher than the external humidity, and the lower the external humidity, the more significant the drying effect, which is reflected in the diffusion coefficient, and the higher the diffusion coefficient value. As can be seen in Figure 4, there is a linear relationship between the moisture diffusion coefficient of concrete and the external humidity. However, there are differences in the ratio between them at different strength classes as well as at different ages. In general, the effect of external humidity on the humidity diffusion coefficient decreases with age and concrete strength.

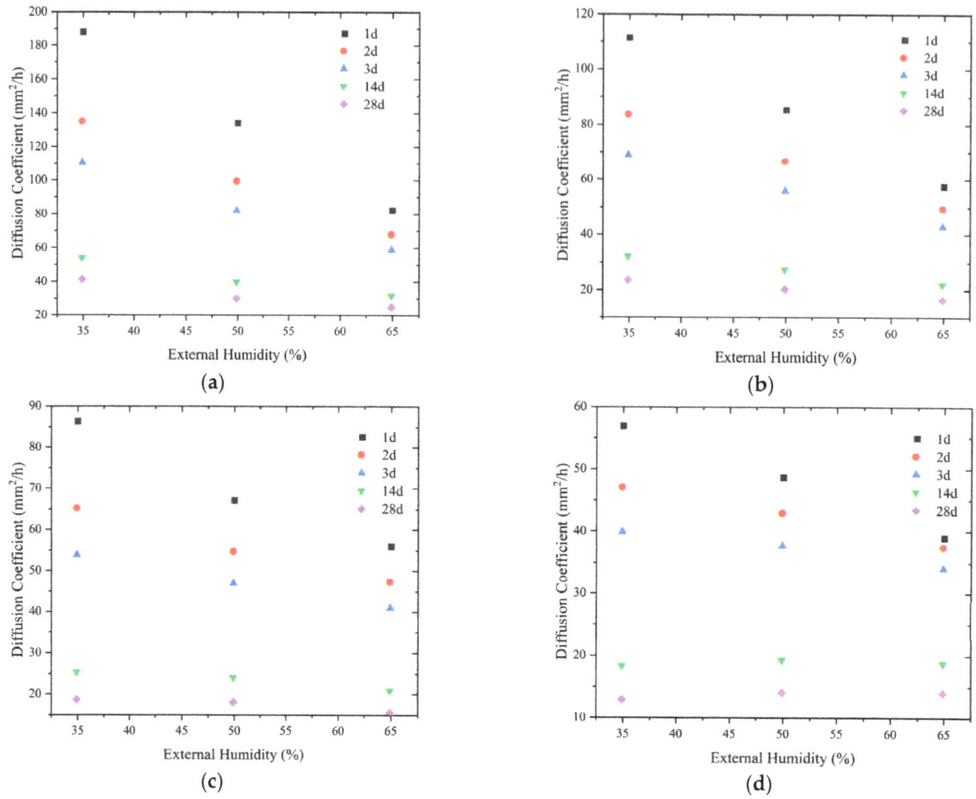

Figure 4. Relationship between moisture diffusion coefficient and external humidity: (**a**) C30; (**b**) C40; (**c**) C50; (**d**) C60.

Figure 5 shows the moisture diffusion coefficients of C30, C40, C50, and C60 concrete for different humidity conditions to elucidate the relationship between the moisture diffusion coefficient of concrete and its internal humidity. Noticeably, a satisfactory correlation exists between the humidity diffusion coefficient of concrete and the relative humidity inside the concrete. The moisture diffusion coefficient of concrete gradually decreases when the humidity decreases. The development of the moisture diffusion coefficient of concrete with humidity can be approximately divided into three stages. When the RH is >95%, the diffusion coefficient decreases rapidly. When the RH is 80–95%, the diffusion coefficient decreases slowly. When the RH is <80%, the reduction rate in the humidity diffusion coefficient decreases significantly. This humidity diffusion coefficient change rule can be explained as follows: when the humidity is >95%, the water content of concrete is higher; meanwhile, the water inside the concrete is mainly liquid water. The diffusion of internal water is also dominated by the evaporation and flow of liquid water, which leads to a higher humidity diffusion coefficient. When the RH is within 80–95%, the water in the concrete changes from mainly liquid water to the liquid water–gaseous water state. During water diffusion, the proportion of gaseous water diffusion gradually increases, which leads to a decrease in the moisture diffusion coefficient. When the RH is <80%, the water in the concrete mainly exists in the gaseous state, and the water content is further decreased, which immensely lowers the value of the diffusion coefficient.

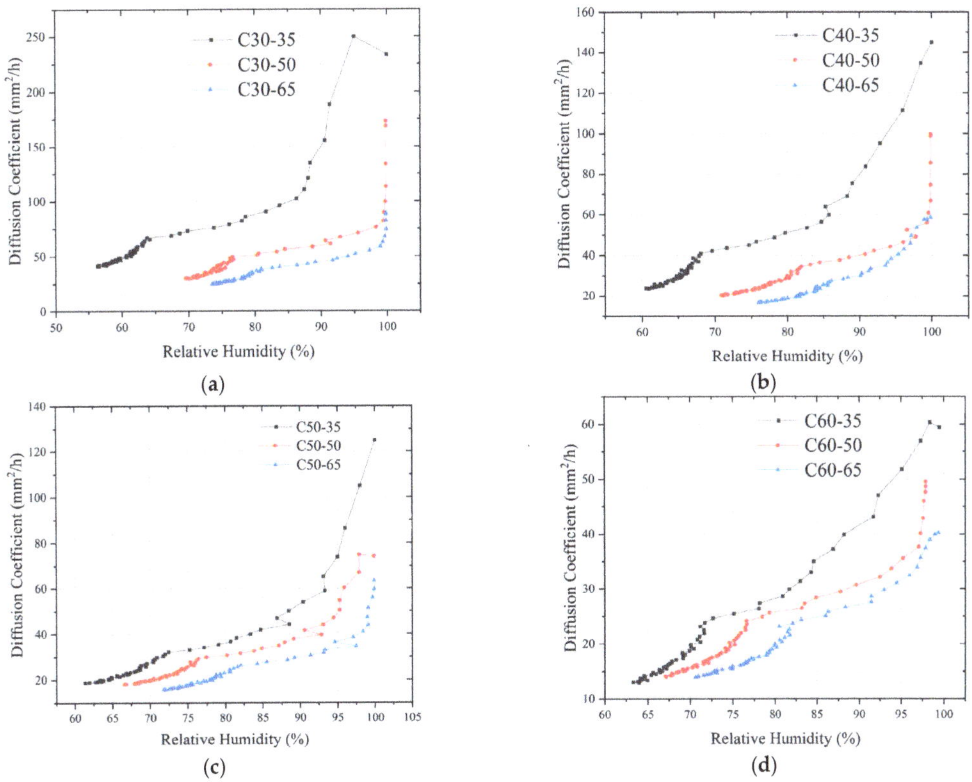

Figure 5. Development of diffusion coefficients of drying components at different internal humidities: (**a**) C30; (**b**) C40; (**c**) C50; (**d**) C60.

After obtaining the moisture diffusion coefficient of the concrete, the internal moisture can be calculated using Equation (29). In this study, the humidity in the middle of the specimen was determined. Therefore, $x = 100$ when calculating the internal moisture. Figures 6 and 7 depict the calculated results of the internal humidity of concrete using Equation (29) as compared to the test results. Notably, these test results refer to the results obtained by subtracting the decrease in humidity due to cement hydration (i.e., the humidity test results of the sealed curing specimens) from the humidity test results of the dry specimens. The comparison results reveal that the calculated models are largely consistent with the test results. The average error over the whole age period was approximately 5%. However, while the computational method has better predictions overall, the overall predictions are not sufficient for earlier ages (up to 2 days). The aforementioned may be related to the distortion of the model, which is attributable to the insufficient understanding of the internal humidity of early-age concrete and the selection of the error function. However, the model can generally predict the internal humidity of concrete satisfactorily, and the error results in an amplification of the internal gradient of concrete due to diffusion; thus, the calculation of concrete shrinkage and shrinkage cracking by the humidity is conservative. Therefore, this type of error is deemed acceptable.

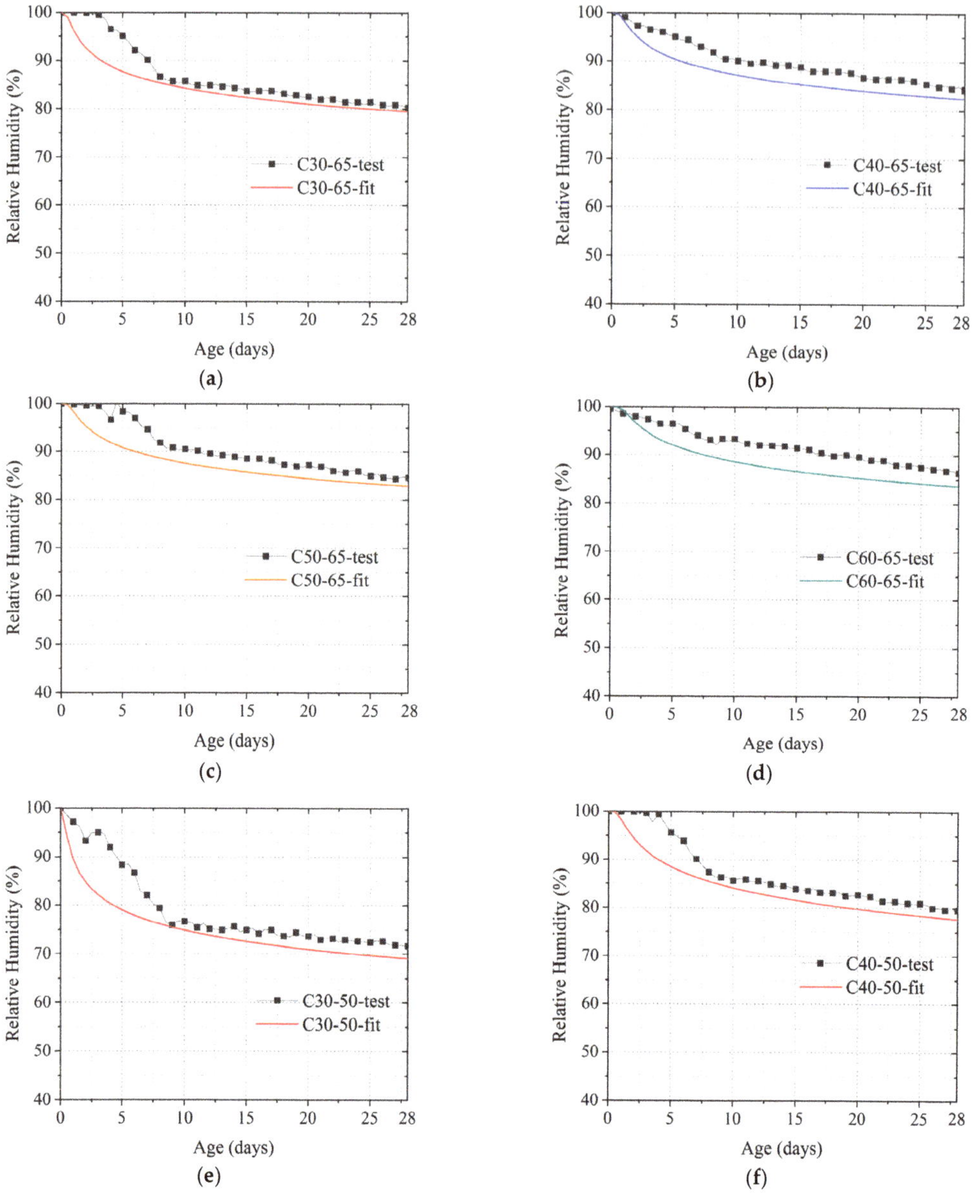

Figure 6. *Cont.*

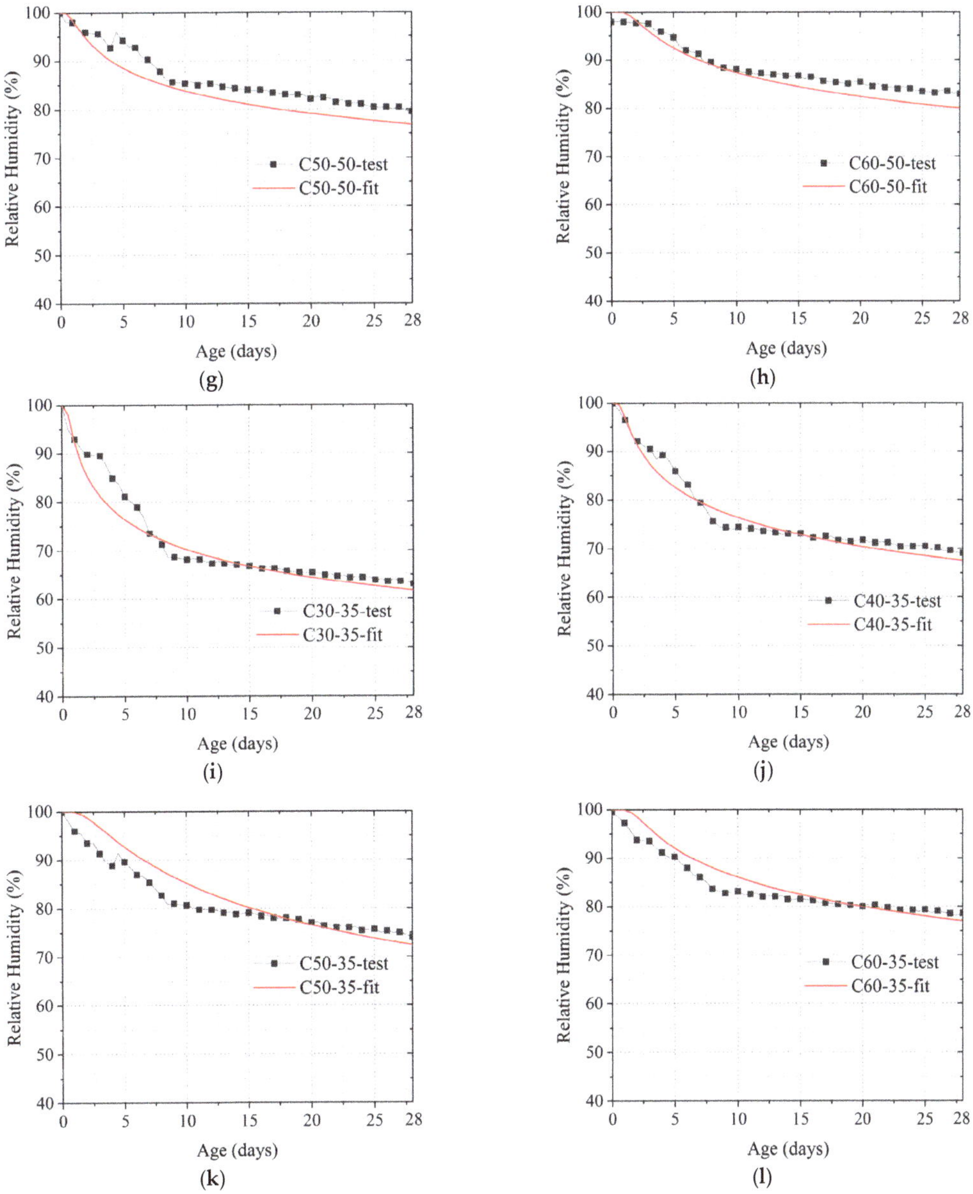

Figure 6. Comparison of calculated and actual internal humidity values of concrete without constraint conditions: (**a**) C30-65; (**b**) C40-65; (**c**) C50-65; (**d**) C60-65; (**e**) C30-50; (**f**) C40-50; (**g**) C50-50; (**h**) C60-50; (**i**) C30-35; (**j**) C40-35; (**k**) C50-35; (**l**) C60-35.

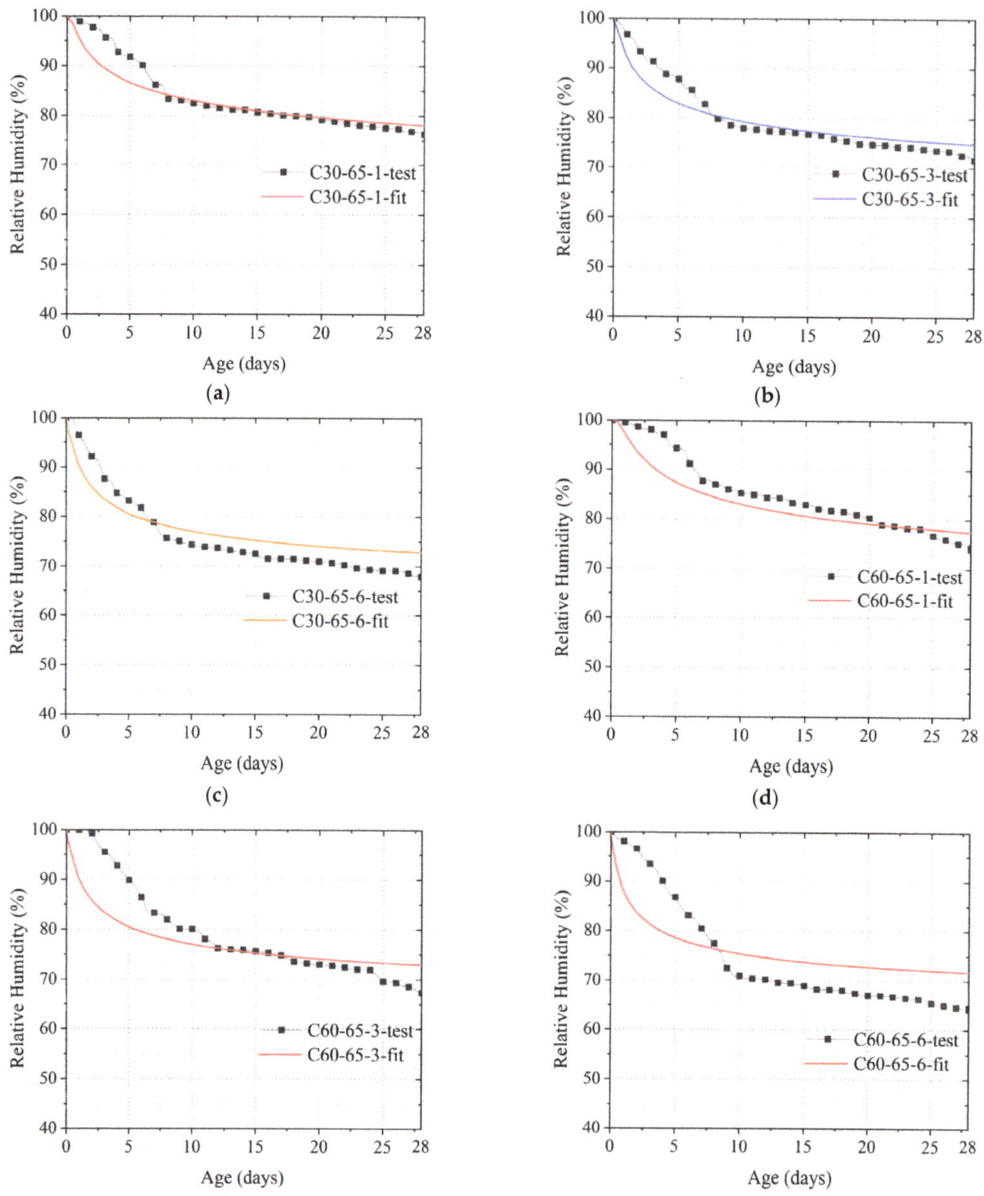

Figure 7. Comparison of calculated and actual internal humidity values of concrete under constraint conditions: (**a**) C30-65-1; (**b**) C30-65-3; (**c**) C30-65-6; (**d**) C60-65-1; (**e**) C60-65-3; (**f**) C60-65-6.

4.3. Effect of Concrete Strength on Moisture Diffusion Coefficient

The effect of different concrete strengths on the diffusion coefficient of humidity for the same external ambient humidity condition is depicted in Figure 8. The diffusion coefficient versus time curves for different concrete strengths under the same external humidity conditions reveal that when the concrete strength is higher, the diffusion coefficient under the same external humidity conditions is lower. In the early concrete-formation stage, this difference is quite apparent. This phenomenon is explained below.

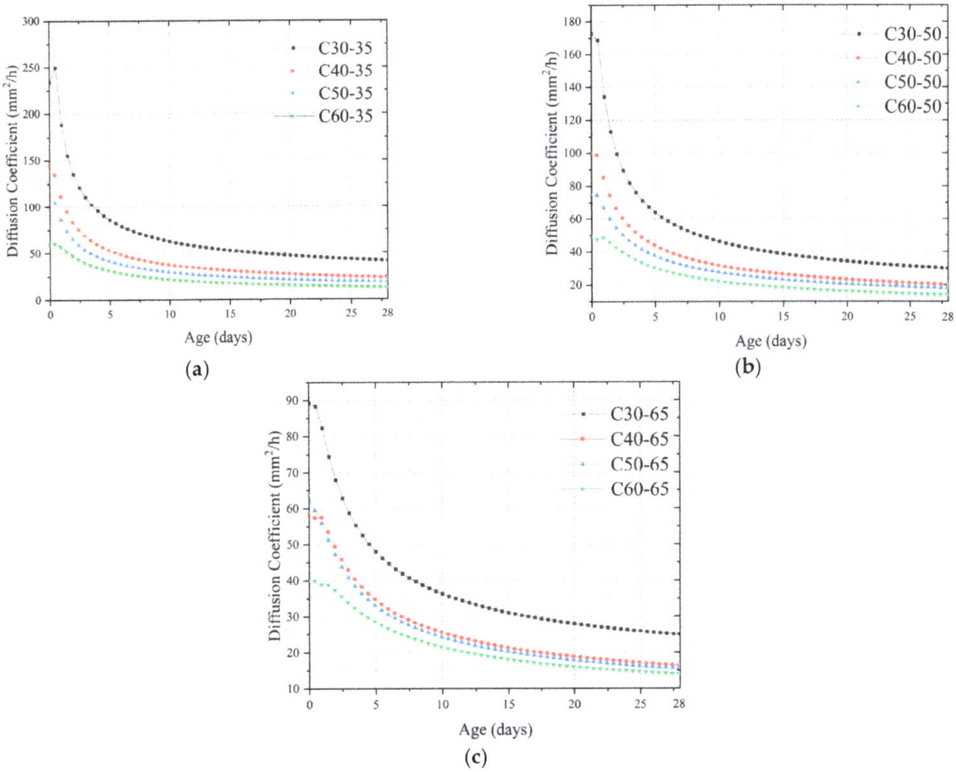

Figure 8. Effect of different concrete strengths on the humidity diffusion coefficient for the same external ambient humidity: (**a**) $RH = 35\%$; (**b**) $RH = 50\%$; (**c**) $RH = 65\%$.

4.4. Effect of Pore Structure on the Moisture Diffusion Coefficient of Concrete

The pore structure parameters were determined via a mercury pressure test. The distribution of the microporous structure and pore characterization parameters for each type of specimen at different ages is depicted in Table 7. The table indicates that in the same specimen, the pore structure parameters of concrete decrease with age. For example, the average pore diameter, middle pore diameter, critical diameter, and porosity of the C30-65 specimen obtained at 3 days were 24.45 nm, 65.17 nm, 61.79 nm, and 22.52%, respectively, which were 5.1 nm, 26.13 nm, 21.29 nm, and 6.20% higher than those at 28 days, respectively. This phenomenon can be explained as follows: with age, the hydration of cement continues to develop, and the hydration products gradually fill the voids in the concrete. Meanwhile, according to capillary tension theory, due to the consumption of water, capillary stress will be generated on the pore walls in the concrete. Under the action of capillary stress, the distance between concrete pore walls and micro-cracks decreases, which significantly contributes to the decrease in pore structure parameters.

Table 7. Pore characteristic parameters of specimens.

	Age (Days)	Average Pore Diameter (nm)	Median Pore Diameter (nm)	Critical Diameter of Capillary (nm)	Porosity (%)	D (mm^2/h)
C30-65	3	24.45	65.17	61.79	22.52	13.21
	7	21.94	46.13	52.30	20.32	10.24
	28	19.30	39.04	40.50	16.32	14.92
C30-65-1	3	26.20	76.10	74.12	25.54	12.96
	7	24.39	52.88	59.49	24.31	11.75
	28	20.58	45.25	52.34	19.21	13.19
C30-65-3	3	30.17	82.02	79.03	27.56	11.22
	7	27.41	64.04	66.60	25.64	11.34
	28	23.42	59.51	55.07	21.61	15.45
C30-65-6	3	32.26	93.27	89.82	29.97	8.88
	7	29.53	78.76	75.43	27.75	13.72
	28	26.35	64.83	62.46	23.84	16.31
C60-65	3	21.35	53.61	44.13	21.52	10.08
	7	17.41	36.53	29.55	18.47	17.80
	28	13.50	30.19	21.23	13.94	28.67
C60-65-1	3	24.87	57.81	53.65	24.83	13.30
	7	22.28	43.07	33.95	21.13	20.58
	28	17.14	36.83	26.78	17.74	17.25
C60-65-3	3	27.85	66.13	57.29	25.94	11.89
	7	24.75	51.46	37.99	22.88	17.49
	28	19.10	39.23	31.00	19.01	19.84
C60-65-6	3	30.01	80.45	68.58	27.92	7.88
	7	27.20	59.36	46.34	25.15	8.59
	28	19.90	45.04	38.91	21.41	12.69

Figure 9 shows the relationship between reinforcement rate and pore structure parameters of concrete at 28 days. Noticeably, the pore structure parameters of concrete increase as the reinforcement ratio increases in concrete specimens of the same strength class. For example, the porosity of the C30-65 (ρ = 0%) concrete specimen at 28 days was 2.89, 5.29, and 7.52% lower than that of C30-65-1 (ρ = 0%), C30-65-3 (ρ = 0%), and C30-65-6 (ρ = 0%), respectively. This phenomenon was also observed in other pore structure parameters as well as in the specimens of the C60 group. This phenomenon is mainly due to the reinforcement hindering the reduction of the pore structure parameters of the concrete due to capillary pore stresses. At 3 days and 7 days of age, the above laws still hold.

When capillary pore stresses are generated in the internal pore walls of concrete due to water loss, the capillary stresses will result in a reduction in the pore wall size. In the restrained specimen, both the capillary stress and the restrained tensile stress generated by the reinforcement act on the pore wall due to the restraining effect of the reinforcement. The stress generated by concrete (σ_{sh-x}) in the restrained specimen can be expressed as follows:

$$\sigma_{sh-x} = \frac{E_s(1-2\mu)}{E_s + E_{re}\rho} \cdot \frac{2\gamma}{a_0 + k\rho} \tag{36}$$

where σ_{sh-x} is the combined capillary pore stress acting on the pore wall, E_s is the modulus of elasticity of concrete, E_{re} is the modulus of elasticity of steel reinforcement, ρ is the reinforcement ratio, μ is the Poisson's ratio of concrete, and a and γ are calculation parameters. The above equation shows that the confining stress of the reinforcement increases with the increase in the reinforcement ratio. In this case, it will decrease the stress ensemble acting on the pore walls of the concrete. This indicates that the extent of decrease in the pore structure parameters of the concrete due to the action of capillary stresses decreases as the reinforcement ratio increases. This agrees with the experimental results that indicate that the pore structure size of the specimens increases with increasing reinforcement rate.

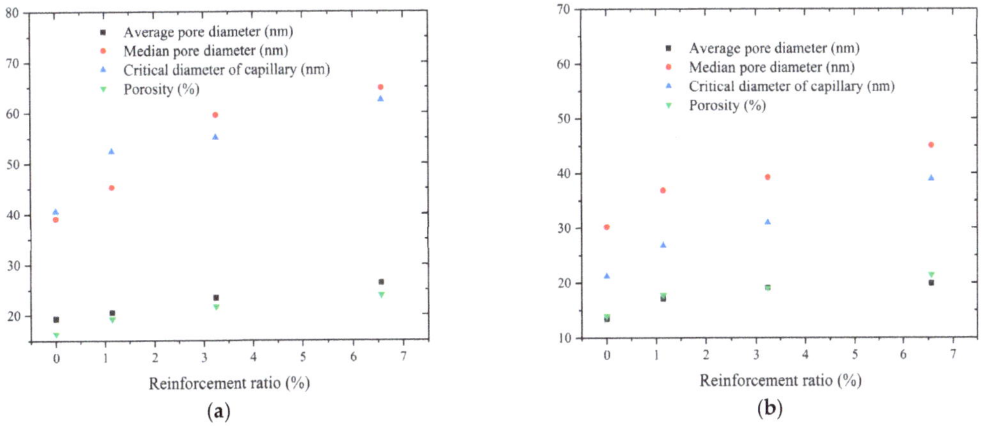

Figure 9. Relationship between reinforcement ratio and pore structure parameters at 28 days: (**a**) C30; (**b**) C60.

Figure 10 shows the relationship between the concrete pore structure parameters and the humidity diffusion coefficient, indicating that the humidity diffusion coefficient increases as the concrete pore structure parameters increase.

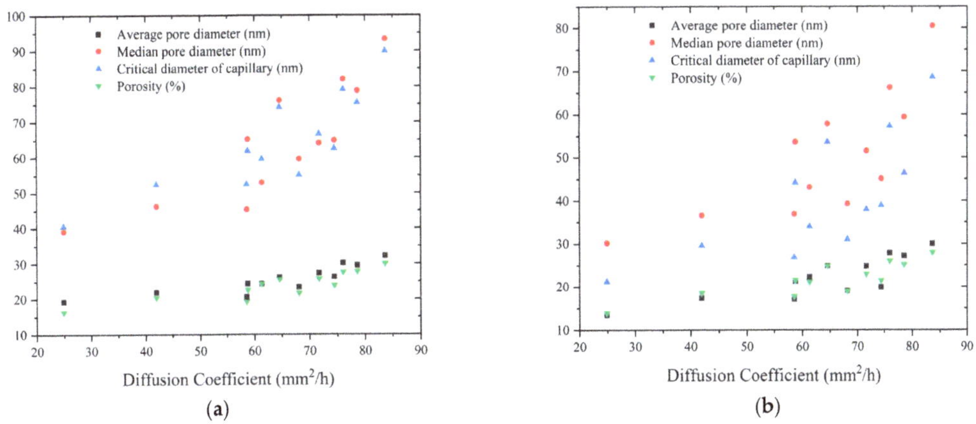

Figure 10. Relationship between concrete pore structure parameters and humidity diffusion coefficient: (**a**) C30; (**b**) C60.

Because pores are the main channel for moisture diffusion in concrete, when the constraint affects the pore structure of concrete, it also affects its moisture diffusion capacity. Figure 11 shows the effect of different constraints (reinforcement ratio) on the moisture diffusion coefficient for the same external ambient humidity and concrete strength. The curve of the moisture diffusion coefficient of concrete with time shows that the higher the internal constraint of concrete, the higher the moisture diffusion coefficient of concrete at the same moment. Taking C30 concrete as an example, the moisture diffusion coefficient of the C30-65 (ρ = 0%) specimen at the age of 3 days was lower than that of C30-65-1 (ρ = 1.14%), C30-65-3 (ρ = 3.24%), and C30-65-3 (ρ = 6.56%) by 5.80 mm^2/h, 34.05 mm^2/h, and 51.13 mm^2/h, respectively.

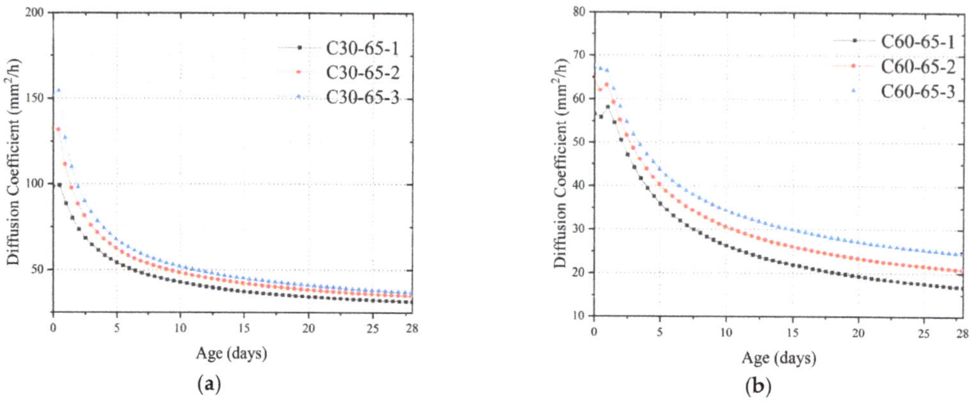

Figure 11. Relationship between constraint conditions and diffusion coefficients: (**a**) C30; (**b**) C60.

5. Conclusions

In this study, a method for calculating the internal humidity of concrete using the concrete humidity diffusion coefficient was theoretically derived. The internal humidity, pore structure parameters, and mechanical properties of concrete under different strength classes, curing conditions, and constraints were tested. The test results were used to investigate the influence laws of these conditions on the concrete humidity and humidity diffusion coefficient.

The following conclusions can be drawn from the results:

1. The internal humidity of concrete varies with age. As the age increases, the internal humidity of concrete decreases and gradually stabilizes. Due to more intense hydration, the internal humidity of high-strength concrete decreases more rapidly under sealed conditions. The internal humidity of specimens under dry conditions decreases faster than that of sealed specimens due to the drying effect. For the same concrete strength class, the lower the external humidity and the higher the reinforcement rate, the faster the internal humidity of the concrete decreases.
2. The moisture diffusion coefficient of concrete can be divided into a rapid decline phase (0–1 day), a slow decline phase (1–8 days), and a stabilization phase (after 8 days) as the age increases. Due to the existence of a humidity gradient, the lower the external environmental humidity, the higher the humidity diffusion coefficient of concrete for the same concrete strength class and reinforcement ratio, and the two exhibit a linear relationship. With the increase in age and concrete strength, the influence of external humidity on the humidity diffusion coefficient of concrete weakens. Meanwhile, along with the gradual decrease in the internal humidity of concrete, the humidity diffusion coefficient of concrete also gradually decreases, and it is more difficult for the water in concrete to be transmitted to the external environment.
3. At the same age, high-strength concrete has smaller pore structure parameters, which makes transporting water in high-strength concrete difficult. As a result, the moisture diffusion coefficient of concrete decreases as the concrete strength class increases for the same external humidity and reinforcement ratio. The method proposed in this study for calculating the internal humidity of concrete using the humidity diffusion coefficient is accurate.
4. Reinforcing bars create confining tensile stresses inside the concrete, which changes the pore structure inside the concrete. For the same concrete strength, the higher the reinforcement rate, the larger the pore structure parameters of concrete (average pore diameter, median pore diameter, critical diameter of capillary, and porosity). Because the pore structure is the main transport channel for water in concrete, the higher the reinforcement ratio, the greater the moisture diffusion coefficient of the concrete and

the faster the moisture in the concrete decreases for the same external humidity and concrete strength.

Author Contributions: F.Z. and W.L.: Conceptualization, Investigation, Methodology, Data curation, Writing—original draft, Writing—review and editing, Formal analysis. L.H.: Supervision, Resources, Project administration, Funding acquisition Writing—original draft, Writing—review and editing, Visualization. Y.H., Z.X., J.Y., D.W. and Z.C.: Writing—original draft, Writing—review and editing. All authors have read and agreed to the published version of the manuscript.

Funding: This research received no external funding.

Data Availability Statement: Not applicable.

Acknowledgments: The authors acknowledge the financial support for publication fees by the School of Civil Engineering, Chongqing University, China. Additionally, the authors are very grateful to Lin Chen for his contribution to this paper.

Conflicts of Interest: The authors declare no conflict of interest.

References

1. Krauss, P.D.; Rogalla, E.A. *Transverse Cracking in Newly Constructed Bridge Decks*; National Academy of Sciences: Washington, DC, USA, 1996.
2. Chen, L.; Huang, L.; Hua, J.; Chen, Z.; Wei, L.; Osman, A.I.; Fawzy, S.; Rooney, D.W.; Dong, L.; Yap, P.-S. Green construction for low-carbon cities: A review. *Environ. Chem. Lett.* **2023**, *21*, 1627–1657. [CrossRef]
3. Zhou, F.; Jiang, H.; Huang, L.; Hu, Y.; Xie, Z.; Zeng, Z.; Liu, M.; Wang, B.; Zhou, X. Early Shrinkage Modeling of Complex Internally Confined Concrete Based on Capillary Tension Theory. *Buildings* **2023**, *13*, 2201. [CrossRef]
4. Huang, L.; Hua, J.; Kang, M.; Zhou, F.; Luo, Q. Capillary tension theory for predicting shrinkage of concrete restrained by reinforcement bar in early age. *Constr. Build. Mater.* **2019**, *210*, 63–70. [CrossRef]
5. Tran, N.P.; Gunasekara, C.; Law, D.W.; Houshyar, S.; Setunge, S.; Cwirzen, A. A critical review on drying shrinkage mitigation strategies in cement-based materials. *J. Build. Eng.* **2021**, *38*, 102210. [CrossRef]
6. Wang, X.; Song, P.; Yu, H.; Taylor, P.; Sadati, S.; Freeseman, K.; Ning, Y. Extended life concrete bridge decks utilizing internal curing to reduce cracking–Materials characterization and engineering demonstration. *Constr. Build. Mater.* **2021**, *275*, 122163. [CrossRef]
7. Miao, Y.; Lu, Z.; Wang, F.; Wang, H.; Li, Y.; Lin, J.; Jiang, J. Shrinkage cracking evolvement in concrete cured under low relative humidity and its relationship with mechanical development. *J. Build. Eng.* **2023**, *72*, 106670. [CrossRef]
8. Yu, S.; Sun, Z.; Qian, W.; Yu, J.; Yang, J. A meshless method for modeling the microscopic drying shrinkage cracking processes of concrete and its applications. *Eng. Fract. Mech.* **2023**, *277*, 109014. [CrossRef]
9. Shen, D.; Kang, J.; Jiao, Y.; Li, M.; Li, C. Effects of different silica fume dosages on early-age behavior and cracking resistance of high strength concrete under restrained condition. *Constr. Build. Mater.* **2020**, *263*, 120218. [CrossRef]
10. Klemczak, B.; Knoppik-Wróbel, A. Reinforced concrete tank walls and bridge abutments: Early-age behaviour, analytic approaches and numerical models. *Eng. Struct.* **2015**, *84*, 233–251. [CrossRef]
11. Chen, L.; Chen, Z.; Zhang, Y.; Liu, Y.; Osman, A.I.; Farghali, M.; Hua, J.; Al-Fatesh, A.; Ihara, I.; Rooney, D.W. Artificial intelligence-based solutions for climate change: A review. *Environ. Chem. Lett.* **2023**, *21*, 2525–2557. [CrossRef]
12. Chen, L.; Msigwa, G.; Yang, M.; Osman, A.I.; Fawzy, S.; Rooney, D.W.; Yap, P.-S. Strategies to achieve a carbon neutral society: A review. *Environ. Chem. Lett.* **2022**, *20*, 2277–2310. [CrossRef]
13. Farghali, M.; Osman, A.I.; Mohamed, I.M.; Chen, Z.; Chen, L.; Ihara, I.; Yap, P.-S.; Rooney, D.W. Strategies to save energy in the context of the energy crisis: A review. *Environ. Chem. Lett.* **2023**, *21*, 2003–2039. [CrossRef] [PubMed]
14. Deysel, R.C.; Boshoff, W.P.; Smit, M.S. Implementing capillary pressure control measures to prevent plastic shrinkage cracking in concrete. *Constr. Build. Mater.* **2023**, *397*, 132407. [CrossRef]
15. Liang, S.; Liu, Y.; Song, G.; Yan, H.; Song, B.; Liu, J. Effect of sulfoaluminate expansive additive on mechanical properties, internal relative humidity, and shrinkage of early-age mortar. *Case Stud. Constr. Mater.* **2023**, *19*, e02226. [CrossRef]
16. Wang, Y.; Zhu, J.; Guo, Y.; Wang, C. Early shrinkage experiment of concrete and the development law of its temperature and humidity field in natural environment. *J. Build. Eng.* **2023**, *63*, 105528. [CrossRef]
17. Persson, B. Moisture in concrete subjected to different kinds of curing. *Mater. Struct.* **1997**, *30*, 533–544. [CrossRef]
18. Kim, J.-K.; Lee, C.-S. Moisture diffusion of concrete considering self-desiccation at early ages. *Cem. Concr. Res.* **1999**, *29*, 1921–1927. [CrossRef]
19. Nilsson, L.-O. Long-term moisture transport in high performance concrete. *Mater. Struct.* **2002**, *35*, 641–649. [CrossRef]

20. Andrade, C.; Sarría, J.; Alonso, C. Relative humidity in the interior of concrete exposed to natural and artificial weathering. *Cem. Concr. Res.* **1999**, *29*, 1249–1259. [CrossRef]
21. Parrott, L. Some effects of cement and curing upon carbonation and reinforcement corrosion in concrete. *Mater. Struct.* **1996**, *29*, 164–173. [CrossRef]
22. Wang, J.; Li, H.; Wang, Z.; Yi, Z.; Huang, F. Humidity field and moisture transfer of concrete with different pre-saturated recycled sand. *Constr. Build. Mater.* **2023**, *382*, 131338. [CrossRef]
23. Shen, D.; Liu, C.; Wang, M.; Jin, X.; Tang, H. Prediction model for internal relative humidity in early-age concrete under different curing humidity conditions. *Constr. Build. Mater.* **2020**, *265*, 119987. [CrossRef]
24. Tazawa, E.-I.; Miyazawa, S. Influence of cement and admixture on autogenous shrinkage of cement paste. *Cem. Concr. Res.* **1995**, *25*, 281–287. [CrossRef]
25. Tazawa, E.-I.; Miyazawa, S. Experimental study on mechanism of autogenous shrinkage of concrete. *Cem. Concr. Res.* **1995**, *25*, 1633–1638. [CrossRef]
26. Aitcin, P.-C. Demystifying autogenous shrinkage. *Concr. Int.* **1999**, *21*, 54–56.
27. Ni, W.-M. *The Mathematics of Diffusion*; SIAM: Philadelphie, PA, USA, 2011.
28. Park, S.-S.; Kwon, S.-J.; Jung, S.H.; Lee, S.-W. Modeling of water permeability in early aged concrete with cracks based on micro pore structure. *Constr. Build. Mater.* **2012**, *27*, 597–604. [CrossRef]
29. Linderoth, O.; Johansson, P.; Wadsö, L. Development of pore structure, moisture sorption and transport properties in fly ash blended cement-based materials. *Constr. Build. Mater.* **2020**, *261*, 120007. [CrossRef]
30. Huang, L.M.; Tang, L.P.; Lofgren, I.; Olsson, N.; Yang, Z.H.; Li, Y.Q. Moisture and ion transport properties in blended pastes and their relation to the refined pore structure. *Cem. Concr. Res.* **2022**, *161*, 106949. [CrossRef]
31. Bažant, Z.; Najjar, L. Nonlinear water diffusion in nonsaturated concrete. *Matériaux Constr.* **1972**, *5*, 3–20. [CrossRef]
32. Ayano, T.; Wittmann, F.H. Drying, moisture distribution, and shrinkage of cement-based materials. *Mater. Struct.* **2002**, *35*, 134–140. [CrossRef]
33. Sakata, K. A study on moisture diffusion in drying and drying shrinkage of concrete. *Cem. Concr. Res.* **1983**, *13*, 216–224. [CrossRef]
34. Akita, H.; Fujiwara, T.; Ozaka, Y. A practical procedure for the analysis of moisture transfer within concrete due to drying. *Mag. Concr. Res.* **1997**, *49*, 129–137. [CrossRef]
35. Ba, M.F.; Qian, C.X.; Gao, G.B. Nonlinear calculation of moisture transport in underground concrete. *Comput. Concr.* **2014**, *13*, 361–375. [CrossRef]
36. Rahimi-Aghdam, S.; Rasoolinejad, M.; Bažant, Z.P. Moisture diffusion in unsaturated self-desiccating concrete with humidity-dependent permeability and nonlinear sorption isotherm. *J. Eng. Mech.* **2019**, *145*, 04019032. [CrossRef]
37. Lin, G.; Liu, Y.; Xiang, Z. Numerical modeling for predicting service life of reinforced concrete structures exposed to chloride environments. *Cem. Concr. Compos.* **2010**, *32*, 571–579. [CrossRef]
38. Wang, A.; Sun, D.; Deng, M.; Chen, X.; Zhang, F. Influence of restraint on pore structures and air permeability of concrete containing MgO-type expansive agent. *Asian J. Chem.* **2013**, *25*, 5532. [CrossRef]
39. Zhang, R.; Wang, Q.; Ma, L. Relations between pore structure and creep characteristics of concrete under micro-expanding and lateral restraints. *J. Chin. Ceram. Soc.* **2015**, *7*, 585–593.
40. Yoo, D.-Y.; Park, J.-J.; Kim, S.-W.; Yoon, Y.-S. Influence of reinforcing bar type on autogenous shrinkage stress and bond behavior of ultra high performance fiber reinforced concrete. *Cem. Concr. Compos.* **2014**, *48*, 150–161. [CrossRef]
41. Sule, M.; van Breugel, K. The effect of reinforcement on early-age cracking due to autogenous shrinkage and thermal effects. *Cem. Concr. Compos.* **2004**, *26*, 581–587. [CrossRef]
42. Huang, L.; Hua, J.; Kang, M.; Zhang, A. Influence of reinforcement configuration on the shrinkage and cracking potential of high-performance concrete. *Constr. Build. Mater.* **2017**, *140*, 20–30. [CrossRef]
43. Yoo, D.-Y.; Park, J.-J.; Kim, S.-W.; Yoon, Y.-S. Early age setting, shrinkage and tensile characteristics of ultra high performance fiber reinforced concrete. *Constr. Build. Mater.* **2013**, *41*, 427–438. [CrossRef]
44. Ming, K. Research on Restrained Deformation Properties of Reinforced Concrete Members due to Early Shrinkage during Construction. Ph.D. Thesis, Chongqing University, Chongqing, China, 2010.
45. Radlinska, A.; Weiss, J. Quantifying variability in assessing the risk of early-age cracking in restrained concrete elements. In *Brittle Matrix Composites 8*; Elsevier: Amsterdam, The Netherlands, 2006; pp. 331–342.
46. Zhang, J.; Huang, Y.; Qi, K.; Gao, Y. Interior relative humidity of normal-and high-strength concrete at early age. *J. Mater. Civ. Eng.* **2012**, *24*, 615–622. [CrossRef]
47. Zhang, J.; Qi, K.; Huang, Y. Calculation of moisture distribution in early-age concrete. *J. Eng. Mech.* **2009**, *135*, 871–880. [CrossRef]
48. Deboodt, T.; Fu, T.; Ideker, J.H. Evaluation of FLWA and SRAs on autogenous deformation and long-term drying shrinkage of high performance concrete. *Constr. Build. Mater.* **2016**, *119*, 53–60. [CrossRef]

49. Shen, D.; Liu, K.; Ji, Y.; Shi, H.; Zhang, J. Early-age residual stress and stress relaxation of high-performance concrete containing fly ash. *Mag. Concr. Res.* **2018**, *70*, 726–738. [CrossRef]
50. *GB/T50081-2002*; Standard for Test Method of Mechanical Properties on Ordinary Concrete. Ministry of Construction of the People's Republic of China and General Administration of Quality Supervision. Inspection and Quarantine of the People's Republic of China: Beijing, China, 2003.

buildings

MDPI

Article

A Quasi-2D Exploration of Mixed-Mode Fracture Propagation in Concrete Semi-Circular Chevron-Notched Disks

Xiaoqing Lu [1,2,*] and Guanxi Yan [2]

[1] Engineering Construction Management Centre, China Nuclear Engineering Consulting Co., Ltd., Beijing 100048, China
[2] School of Civil Engineering, The University of Queensland, St Lucia, QLD 4072, Australia
* Correspondence: xiaoqing.lu1988@outlook.com

Abstract: Most semi-circular bend (SCB) tests on concrete have been conducted with a pre-crack with a straight-through tip, thereby undermining the determination of the tensile fracture toughness (K_{Ic}). Therefore, the present study involved mixed-mode (tensile–shearing) fracture propagation in concrete semi-circular chevron-notched disks (i.e., with a sharp notch tip) using SCB tests and the FRANC2D numerical simulation software. The inclined notch angle (β) was varied from 0° to 70° while the other settings remained fixed, and the crack mouth opening displacement (CMOD) of the notch was measured constantly. The stress distribution was analyzed using finite-element simulations, and the experimental results showed that this testing method was robust. The maximum failure load and the fracture propagation angle increased with β, and wing fracture was observed. With FRANC2D simulating these SCB tests successfully, it was found that the tensile stress concentration around the notch tip moved toward the upper face of the notch, and the compressive stress concentration formed on the notch tip. The tensile mode was generated as the CMOD kept increasing for β = 0–30°, whereas the mixed mode became more evident as the CMOD kept decreasing for β = 45–70°. The fracture process zone was found for β = 0–30° but not for β = 45–70°. This mixed-mode fracture is predicted better by the criterion of extended maximum tangential strain than by other criteria, and there is a linear relationship between CMOD and K_{Ic}, as examined previously for pavement and concrete materials.

Keywords: SCB test; concrete SCCND; FRANC2D; mixed-mode fracture; fracturing angle; CMOD; fracture toughness

1. Introduction

Brittle materials such as glass, rock, and concrete are essential construction materials in civil engineering. Because their mechanical properties raise concerns about the safety of designs in structural and geotechnical engineering, such materials have long been investigated. Initially, because of the success of applying linear elastic theory to ductile materials, the stress–strain constitutive relationship of brittle materials was treated as a linear elastic without deeply investigating the heterogeneity induced by pre-existing fractures and cracks. Compared to ductile materials, the mechanical properties of brittle materials were less well-known and understood in terms of irreversible strain resulting from material fracturing [1]. From a microscale perspective, crack propagation in rock and concrete is a process of atomic decohesion caused by mechanical instability [2]. Because of cyclic loading or environmental conditions, material cracks can be generated during manufacturing and construction. Cracked concrete and rock have posed dangers to and shortened the life spans of many civil infrastructures, so more research is justified to accurately characterize the mechanical performance of such materials under fracturing conditions.

The first to investigate brittle materials at the macroscale was Griffith [3], who developed the theory of energy decohesion in fracture mechanics based on the strain energy release rate during crack propagation. In 1939, Weibull introduced the statistical distribution of initial defects and gave a failure model as a function of stress and volume [4]. Since then, various models of quasi-brittle materials have been established based on initial micro-defects [2]. Meanwhile, Barenblatt [5] and Dugdale [6] devoted their efforts to integrating cohesive forces in the crack tip region within the limitations of the elastic theory. Because the stress conditions near the cracking area vary significantly after crack initiation, cracking patterns are crucial for analyzing fracturing in concrete and rock. According to the theory of linear elastic fracture mechanics, the cracking process begins if the critical stress intensity factor (SIF)—well known as the fracture toughness (K_c)—is reached ($K \geq K_c$). The SIF (K) can be quantified based on the fracture mode near the cracking region, the crack inclination angle, and the crack geometry. Figure 1A shows that crack propagation varies with crack geometry (e.g., elliptical or with zero opening width) and orientation (differing inclination angle). Also, based on the stress conditions around the crack and the crack surface displacement [4,7,8]. There are three fundamental fracture modes: tensile (mode I), shearing (mode II), and tearing (mode III). As shown schematically in Figure 1B, mode I is a tensile mode that involves the two cracking faces separating along the y–axis perpendicular to the x–z plane, mode II is a shearing mode that involves the two cracking faces in the x–z plane sliding in parallel along the x–axis, and mode III is a tearing mode that involves the two cracking faces in the x–z plane sliding along the z–axis.

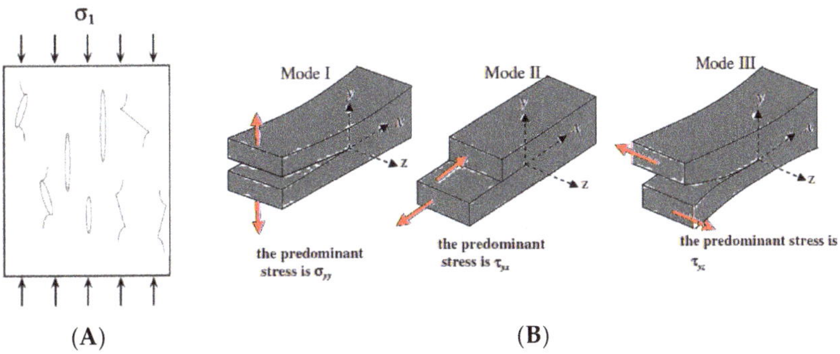

Figure 1. Influential factors for stress intensity factor (SIF): (**A**) examples of crack geometry and orientation; (**B**) three fundamental fracture modes.

Several theories have been established to elucidate the fracturing behavior, including the maximum strain energy release rate criterion (G criterion), the minimum strain energy density criterion (S criterion), the maximum principal stress/strain criterion, the maximum tangential stress (MTS) criterion, and other empirical criteria [4,7]. Compared to principal-stress-based criteria that are applicable only to mode I, energy-based criteria are more suitable for mode II because they can consider shear-stress–strain behavior rather than just principal stress [9]. However, Bocca, et al. [10] noted that neither the G criterion nor the S criterion could be applied straightforwardly for a mixed mode (tensile–shearing). Therefore, Sih [11] and Shen and Stephansson [12] established the modified G criterion, also known as the F criterion, which is a general form of the G criterion considering both modes I and II. Subsequently, to explain mixed-mode fracturing, Lim and Lee [13] developed the generalized MTS (GMTS) criterion based on the F criterion, thereby characterizing K_c mathematically for modes I and II (K_{Ic} and K_{IIc}) and enabling K_c to be quantified for mixed-mode fracturing.

Much research effort has gone into studying concrete fracturing behavior thoroughly under mode-I loading conditions [14]. However, in practice, cracked structures sometimes experience mixed-mode loading combining modes I and II [15]. Also, Shi [16] noted that most fracture problems in concrete are mixed-mode ones involving modes I and II. Therefore, many semi-circular bend (SCB) tests have been conducted under mixed-mode loading conditions by applying a straight-through notch to a pre-cracked concrete specimen [14,17–20]. However, the cracked chevron notched Brazilian disk (CCNBD) test from rock mechanics shows that a sharp cracking tip contributes to more accurately quantified fracturing behavior than does a straight-through notch because the cracking-tip bluntness of the latter undermines the determination of K_c [14]. Nevertheless, despite some previous studies of concrete having tested semi-circular chevron-notched disks (SCCNDs), SCCND concrete specimens have rarely been investigated under mixed-mode loading conditions. Therefore, the present study used experimental and numerical methods to explore mixed-mode (tensile–shearing) concrete crack propagation in SCCND specimens with different chevron notch angles. The study had five aims: (i) assess whether the present experiment is reliable and robust for testing concrete; (ii) study how the crack initiation angle varies with the chevron notch angle; (iii) investigate the crack mouth opening displacement (CMOD) under mixed-mode fracturing; (iv) analyze fracturing behavior with K_c determined by the GMTS criterion; (v) model the crack propagation in an SCCND concrete specimen via numerical simulation. The expectation is that via this series of analyses, the present experimental and numerical work will advance the understanding of mixed-mode (tensile–shearing) crack propagation in concrete SCCND specimens with a chevron notch with a sharp cracking tip so that the concrete fracturing tests can be improved accordingly in any newly updated testing standards in consideration of chevron notch instead of the straight-through notch.

2. Experimental and Numerical Methods

2.1. Specification of Concrete SCCND

All the specimens were prepared according to ISRM (International Society for Rock Mechanics) standards. Following the instructions provided by BASF Ltd. of New Zealand, the concrete samples were fabricated by mixing 20 kg per bag of MasterEmaco N 102 Cl mortar with 3.8–4.0 L of water. Each sample was loaded into a CCNBD mold and then removed after initial concrete curing for four days; note that molding is always better than cutting for generating the chevron notch because the concrete aggregates around the chevron notch lead to poor cutting performance. After curing for 28 days, Young's elastic modulus (E) and the compressive strength reached 10 GPa and 25 MPa, respectively, with a Poisson's ratio (v) of 0.2 [21]. Each CCNBD specimen was split precisely into two equivalent SCCND specimens using a diamond saw.

An SCCND specimen with a chevron notch therein is illustrated in Figure 2A. Its radius (R) was 51.5 mm, its thickness (B) was 30 mm, the chevron notch thickness (T) was 2 mm, the chevron notch angle (β) was 0°, 30°, 45°, 60°, or 70°, the inner chevron notch length (a_0) was 12.5 mm, the outer chevron notch length (a_1) was 23 mm, the chevron notch length (a) was between a_0 and a_1, and the manufacturing standard deviation (STD = $\pm$1 mm) was determined using a caliper. As shown in Figure 2B–D, there were three testing groups: group 1 had β = 0° and 30°, group 2 had β = 0°, 30°, 45°, and 70°, and group 3 had β = 0°, 30°, 45°, and 60°. Because each CCNBD specimen provided two SCCND specimens, each SCB test could be repeated twice for each value of β.

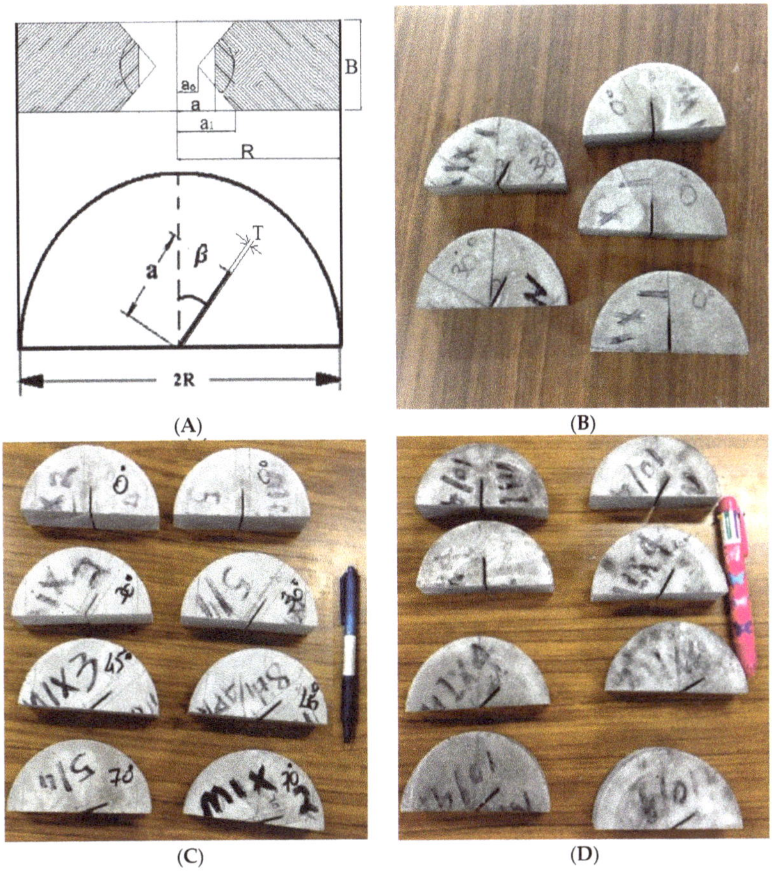

Figure 2. Concrete semi-circular chevron-notched disk (SCCND) specimens with various chevron notch angles (β): (**A**) schematic of SCCND specimen and chevron notch therein; (**B**) testing group 1 ($\beta = 0°, 30°$); (**C**) testing group 2 ($\beta = 0°, 30°, 45°, 70°$); (**D**) testing group 3 ($\beta = 0°, 30°, 45°, 60°$).

2.2. SCB Tests for Concrete SCCND

The SCB test is shown schematically in Figure 3A. It is a three-point bending test with two symmetrical rollers supporting an SCCND specimen when applying a compressive top load P. The ISRM standards suggest that the span between the two rollers ($2S$) should be such that $0.5 < S/R < 0.8$ [22], and they recommend $S/R = 0.6$ for rock SCCND tests ($S = 30$ mm) and $S/R = 0.8$ for concrete ones ($S = 41.5$ mm). Figure 3B shows a universal loading machine (Instron model 4505) with a three-point bend loading frame (i.e., flexural bend fixture). This machine was equipped with a 100-kN load cell, and the hydraulic servo-control testing system was controlled by the Bluehill 3 software. The vertical displacement was measured using a flexural bend fixture with a 100-mm length scale installed in the Instron loading frame. This loading frame applied the top load P along the SCCND central axis by a top roller. Note that mislocating this roller could lead to eccentric loading, so according to ISRM standards, the test specimen should be covered with strips of adhesive paper for better contact with the top roller [22]; alternatively, a potentially better solution is a loading plate, as shown in Figure 3C. The loading condition was controlled in the range of 0.06–0.1 kN (depending on the practical situation) to achieve a displacement rate of 0.025 mm/min, which was computer-controlled by the software Bluehill 3 of Instron model 4505. This loading setting ensured the quasi-static loading conditions and avoided any possible loading-induced mechanical impact by setting an irrationally fixed loading rate. Meanwhile, the two bottom rollers allowed the two segments of a cracked SCCND to rotate around them during a fracturing process.

(A) (B) (C)

Figure 3. Semi-circular bend (SCB) test: (**A**) schematic; (**B**) photograph of a three-point bend loading frame (i.e., flexural bend fixture); (**C**) measurement of crack mouth opening displacement (CMOD).

Figure 3C shows the CMOD measurement in the case of the alternative top-loading method. The CMOD was measured using a crack opening displacement (COD) transducer (Instron model 2670-116) that had a clip gauge attached to the surfaces of the two segments of an SCCND by a binding agent (Loctite 454 instant adhesive). The gap of the clip gauge was initially set to 10 mm and then expanded with subsequent fracturing; thus, the transducer constantly detected the displacement of the clip gauge, which was equal to the CMOD. Also, to investigate the relationship between the chevron-notch angle and the subsequent fracturing angle, all cracking angles were measured using a protractor. Each SCCND specimen was tested following the testing procedure recommended in the ISRM standards [22]. Each test continued until a loading-induced failure occurred, and the failure load (P_{max}), vertical displacement, and CMOD were recorded continually using a computerized data logger to quantify the fracturing behavior.

2.3. Numerical Simulation of SCB Tests

As mentioned above, an SCB test provides three designated variables at the testing scale but cannot detect the stress distribution in an SCCND specimen. Therefore, numerical simulations based on the finite-element method (FEM) were carried out to replicate the present SCB tests under the same testing conditions while also studying the stress distribution in the concrete SCCND. Initially developed by Wawrzynek and Ingraffea [23] at Cornell University, the FRANC2D (Fracture Analysis Code 2D) software was used to conduct these numerical simulations. FRANC2D integrates FEM analysis with fracture mechanics to provide versatile functionalities, including automatic mesh generation, a topological database for local re-meshing [24], elastic analysis with stress singularity [24], computer graphics, and finite-element post-processing [23]. Therefore, academics working on solid and rock mechanics have used FRANC2D extensively over the past two decades to model mixed-mode (tensile–shearing) fracturing [14,25–27].

In the FRANC2D numerical models, the quasi-2D SCB tests were treated as a plane stress problem, as shown in Figure 4. Consequently, the specimen geometry simulated by FRANC2D was the same 2D geometry as that in the actual SCB tests regarding the specimen radius (R), the half-span of the flexural bend fixture (S), the chevron notch length (a), the chevron notch thickness (T), and the chevron notch angle (β), and the specimen with this geometry was assumed to be a continuous, isotropic, and homogeneous elastic specimen. The SCCND material properties were set as E = 10 GPa and v = 0.2 to replicate the aforementioned concrete [21]. Each simulated specimen was meshed using the CASCA automatic mesh generator before the three-point loading simulation (see Figure 4); the mesh resolution at the semi-circular top boundary and the straight-edged bottom boundary was ca. 50 elements, while there were ca. 30 elements around the chevron notch in each simulated SCCND, based on the previous simulating experience using FRANC2D for similar concrete specimens [14]. In Figure 4, the segments on either side of the chevron notch are designated as the "upper face" (left) and the "lower face" (right), and the top end of the chevron notch is the cracking "tip" where the fracture is initiated.

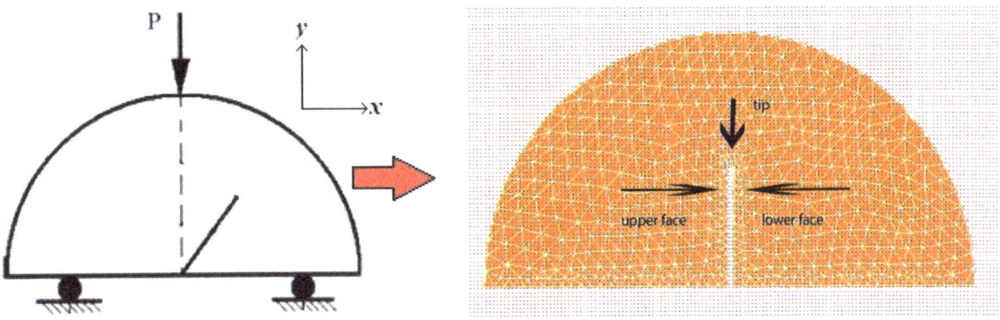

Figure 4. Simulation setup of SCB test in FRANC2D (**left**) and mesh geometry by CASCA (**right**).

In each FRANC2D simulation, kinematic constraints were applied simultaneously to the SCCND specimen at its top, bottom left, and bottom right to fix it in the x and y directions. The three-point bending load in an SCB test was simulated as a uniformly distributed load within an infinitesimal length in the actual 3D space, which could be seen as a point-wised load in a quasi-2D plane. Also, the maximum top load along the central axis and on the two symmetrical supports under the specimen was the failure load obtained from the actual SCB test. The loading conditions were modelled under the quasi-static loading conditions, replicating the practical SCB tests mentioned above. For $\beta = 0°$ and 30°, the cohesive fracture model was applied because of the occurrence of only the tensile mode (type I). For the other values of β at which the tensile–shearing mixed mode (types I and II) occurred, automatic noncohesive fracture propagation was implemented in the FRANC2D program to model the fracture propagation accurately.

Before fracture was initiated, contour plots of the tensile and shear stresses were produced to show the stress distributions in the 2D specimens simulated by FRANC2D. In addition, the normal and shear stress distributions on the upper and lower faces were plotted along position lines using "line plot" in the "post-process" option of FRANC2D. After the simulations, the simulation outputs for the fracture propagation were compared with snapshots of the actual fracture propagations in the SCB tests. These model outcomes could be used to analyze the fracture initiation and propagation, the stress distribution in the SCCND specimen, and the fracturing behavior under mixed-mode loading (β = 0–70°). Therefore, FRANC2D is a useful and powerful software for fracture analysis to aid the standard and modified SCB tests for brittle materials, including glass, rock, and concrete.

2.4. Methods for Fracture Analysis

To analyze the fracture propagation for various values of β in the range of 0–70°, the GMTS criterion was selected to determine K for the mixed mode (types I and II). Then, K could be given separately as the tensile SIF (K_I) for mode I and the shear SIF (K_{II}) for mode II [28,29]. For compatibility with the geometry of an SCCND specimen, Kuruppu and Chong [30] modified the GMTS criterion based on the ISRM standards to obtain

$$K_I = \frac{P_{\max}\sqrt{\pi a}}{2RB} Y_I(\beta, \frac{S}{R}, \frac{a}{R}) \tag{1}$$

$$K_{II} = \frac{P_{\max}\sqrt{\pi a}}{2RB} Y_{II}(\beta, \frac{S}{R}, \frac{a}{R}), \tag{2}$$

where Y_I and Y_{II} are the nondimensional SIFs for modes I and II, respectively, depending on β, S/R, and a/R, and the variables are as follows: $P_{\max}$ is the failure load applied along the central axis of the SCCND specimen; a is the length of the chevron notch in the specimen; S is the half-span of the symmetric supports (i.e., flexural bend fixture) under the specimen; R is the specimen radius; B is the specimen thickness. The relationships between these nondimensional SIFs (Y_I and Y_{II}) and the three parameters β, S/R, and a/R have been given by Ayatollahi and Aliha [31]. In Figure 5, Y_I and Y_{II} were determined using interpolation between the settings of a/R = 0.4 and 0.5 for the same S/R = 0.8 from the comprehensive datasets for Y_I and Y_{II} provided by Ayatollahi and Aliha [31] and Luo, et al. [21], who numerically explored a series of SCB test settings in ranges including those of the present concrete SCCND specimens (i.e., β = 0–70°, S/R = 41.5 mm ÷ 51.5 mm = 0.8, and a/R = 23 mm ÷ 51.5 mm = 0.45).

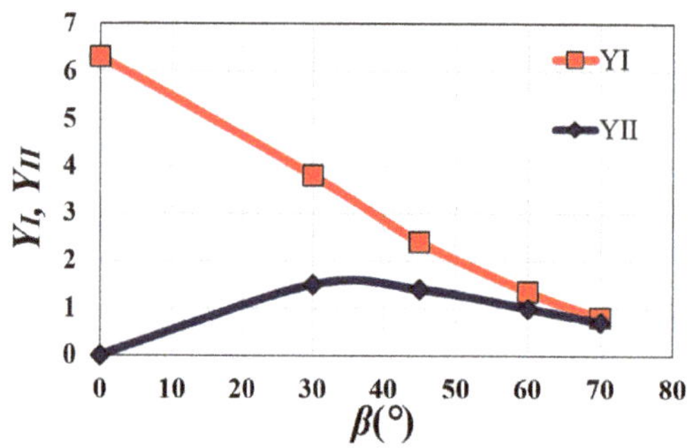

Figure 5. Variations of Y_I and Y_{II} with β = 0–70° for S/R = 0.8 and a/R = 0.45; Y_I and Y_{II} were calculated using interpolation between the settings of a/R = 0.4 and 0.5 for the same S/R = 0.8 from the comprehensive datasets for Y_I and Y_{II} provided by Ayatollahi and Aliha [31].

The critical SIF (K_c) of a brittle material—usually called the fracture toughness—is also characterized by the energy per unit area required to create new crack surfaces, thereby propagating a crack through the material, and it has been identified as the fundamental parameter for understanding the behavior of fractures [32]. K_c reflects the ability of a material to absorb energy, and it is usually determined under impact loading conditions in the presence of a chevron notch, such as in CCNBD and SCB tests. Mixed-mode (tensile–shearing) fracturing can be provided by varying β within the range of 0–90° [33] and the ratios a/R and S/R [28]. In the present work, the mode-I fracture toughness K_{Ic} is given as

$$K_{Ic} = \frac{P_{\max}\sqrt{\pi a}}{2RB} Y_{Ic}, \tag{3}$$

$$Y_{Ic} = -1.297 + 9.516\frac{S}{R} - (0.47 + 16.457\frac{S}{R})\frac{a}{R} + (1.071 + 34.401\frac{S}{R})(\frac{a}{R})^2, \tag{4}$$

where Y_{Ic} is the nondimensional SIF derived under plane stress conditions in a series of FEM analyses by Kuruppu, et al. [22].

Furthermore, based on the weight-function method developed by Adamson, et al. [34], Mubaraki and Sallam [35] gave the relationship between the CMOD and K_I as

$$\text{CMOD}{\cdot}E'\frac{\sqrt{R}}{a} = k{\cdot}K_I + b, \tag{5}$$

where the variables are as follows: E' is the generalized Young's elastic modulus of a brittle material, which is equal to E for plane stress and $E/(1-v^2)$ for plane strain, where v is the Poisson's ratio of the material; R is the radius of the SCCND specimen; a is the length of the chevron notch; k and b are fitting parameters. Whether this linear relationship is valid for a concrete SCCND specimen was also investigated with the experimental measurement of the CMOD and the calculated K_I.

3. Results and Discussion

3.1. Maximum Failure Load

The values of P_{max} on the concrete SCCND specimens were obtained from the SCB tests by following the aforementioned experimental method; they are summarized in Table 1, and the mean values and standard deviations (STD) of P_{max} are plotted in Figure 6. As can be seen, the P_{max} STD changes little with β, thereby showing the reliability and robustness of the present SCB tests. In contrast, P_{max} clearly increases with increasing β, and the reason for this could be that higher shear stress occurs with higher β (45–70°). An example that could explain this situation is that the normal stress (tensile and compressive) applied on the chevron notch tip may effectively serve as shear stress along the original plane of the chevron notch between the upper and lower faces. Meanwhile, almost no shear stress acts on the chevron notch for β = 0–30° because now the normal stress serves mainly as tensile stress along the original plane of the chevron notch between the upper and lower faces. Consequently, there is a lower value of P_{max} for β = 0–30° compared to that for β = 45–70°. The inconsistency of P_{max} > 1.11 kN for β = 60° while P_{max} < 0.98 kN for β = 70° might have one or more causes, including human operating error, the eccentricity of the top loading point, local heterogeneity in the concrete specimens (the embedded fine and coarse aggregates), the mechanical tests being quasi-2D (actually 3D with a thickness B = 30 mm). On the other hand, it might additionally be attributed to P_{max} almost approaching its maximum value when β reaches ca. 60–70°; meanwhile, the sensitivity of β variation was already within the testing and measuring precision.

Table 1. Maximum failure load (P_{max}) on concrete SCCND specimens manufactured by splitting each CCNBD specimen equally into first and second half CCNBD specimens.

β	P_{max} [kN]	Group 1	Group 2	Group 3	Mean	STD
0°	First half CCNBD	0.687	0.211	0.529	0.483	0.232
	Second half CCNBD	0.767	0.141	0.563		
30°	First half CCNBD	0.663	0.520	0.193	0.571	0.182
	Second half CCNBD	0.746	0.670	0.632		
45°	First half CCNBD	–	0.740	0.888	0.867	0.098
	Second half CCNBD	–	0.830	1.010		
60°	First half CCNBD	–	–	1.011	1.112	0.101
	Second half CCNBD	–	–	1.213		
70°	First half CCNBD	–	1.185	–	0.983	0.203
	Second half CCNBD	–	0.780	–		

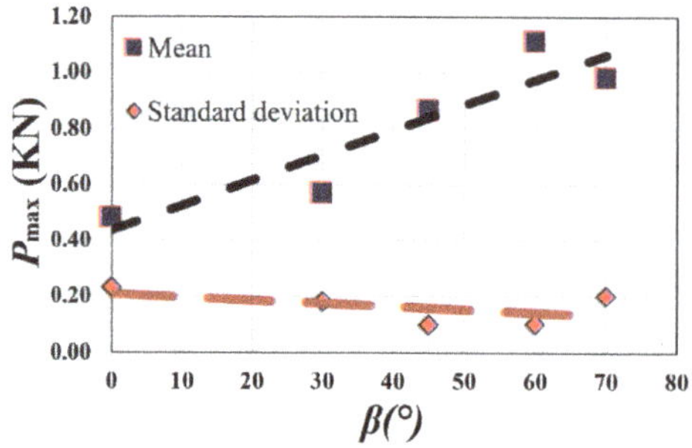

Figure 6. Variations of mean value and standard deviation of P_{max} with β.

3.2. Chevron-Notch and Fracturing Angles

After the SCB tests, the three testing groups of the prepared SCCND specimens in Figure 2B–D were cracked, as shown in Figure 7A–C, respectively. As well as the prefabricated chevron notch angles (β), the fracture propagation angles (θ) were also measured using a protractor and are summarized in Table 2.

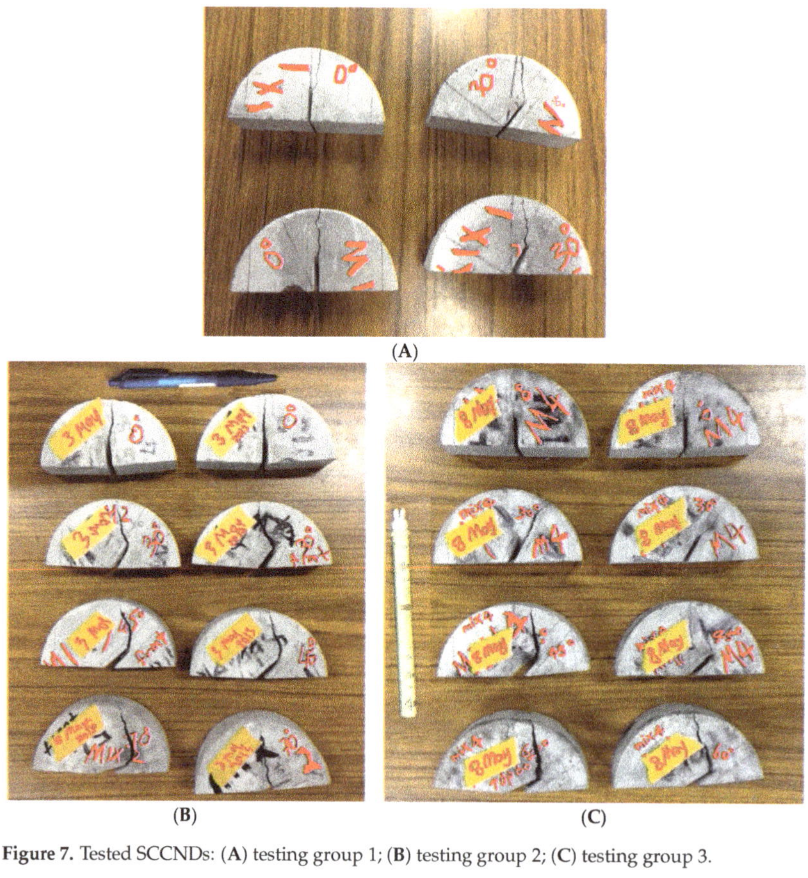

Figure 7. Tested SCCNDs: (**A**) testing group 1; (**B**) testing group 2; (**C**) testing group 3.

Table 2. Values of β and θ for all three testing groups.

β	Fracture Propagation Angle (θ)				
	Group 1	Group 2	Group 3	Mean	STD
0°	0–10°	0–5°	2–5°	4°	3°
30°	30–45°	55–68°	5–15°	36°	22°
45°	–	38–57°	55–65°	54°	10°
60°	–	–	78–80°	79°	1°
70°	–	44–96°	80–96°	79°	21°

To show better the relationship between β and θ, the mean value and STD of the measured values of θ are plotted against β in Figure 8. From these experimental results, it is apparent that θ depends strongly on β given the monotonically increasing relationship between the mean values of β and θ shown in Figure 8. For $\beta \leq 30°$, the tensile-stress-induced fracture propagated at the tip of the chevron notch only and also perpendicular to the two faces of the SCCND specimen. For $\beta > 30°$, the original location of fracture propagation tended to move to the upper face of the chevron notch instead of its tip. In addition, the fracture propagation deviated from the original notch angle of β, and the fracture then propagated to the central loading point by following a curvilinear trajectory (see the demonstrations in Figure 1A and the actual instances in Figure 7B,C). Because these fracturing patterns look like a wing attached to the tip of the chevron notch, they are called wing fractures. This type of fracture is caused by both tensile and shear stress and, therefore, is identified as a mixed-mode (tensile–shearing) fracture. Moreover, note that based on the criterion of maximum tangential stress, the pure shearing mode (type II) loading condition could be generated if $\beta \geq 70.53°$ [29], but this mode was beyond the scope of the present study because of selecting β from the range of 0–70°.

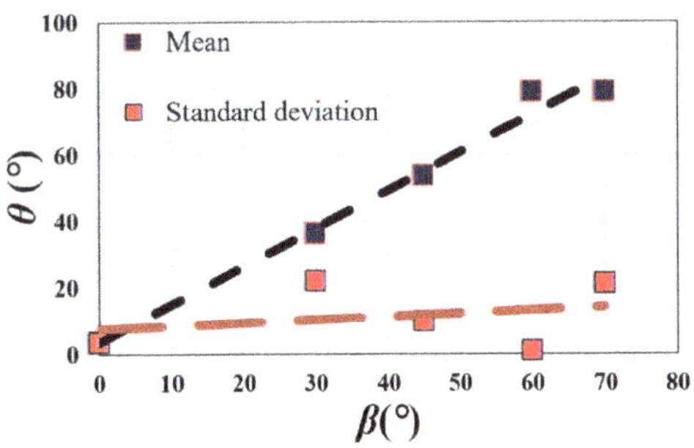

Figure 8. Variations of mean value and standard deviation of θ with β.

Besides, it is clear to see that the STDs of $\beta = 30°$ and $\beta = 70°$ are over 20°, which is relatively significant to undermine the reliability of SCB tests. However, it is worth noting that SCB tests on concrete and rock samples always encounter any possible human operating error, the eccentricity of the top loading point, and local heterogeneity in the concrete specimens (the embedded fine and coarse aggregates). Such uncertainties could potentially cause such significant STDs, quantifying their maximum imprecision. In addition, the ranges of θ were provided because each CCNBD can provide two SCCND, as previously mentioned.

3.3. Stress Distributions in SCCND by FRANC2D

In addition to the mechanical SCB tests, numerical simulations of these tests were carried out using FRANC2D to show the stress distributions in the concrete SCCND specimens. This was done because the mechanical tests were incapable of delivering such outcomes. As can be seen from Figure 9, a common feature among all the simulated stress distributions was that there were similar stress concentrations of compression (highlighted in orange) in the regions adjacent to the three loading points of the SCB tests. In addition, the stress distributions close to those three regions were more prone to compression, whereas those farther away were more tensile (highlighted in white).

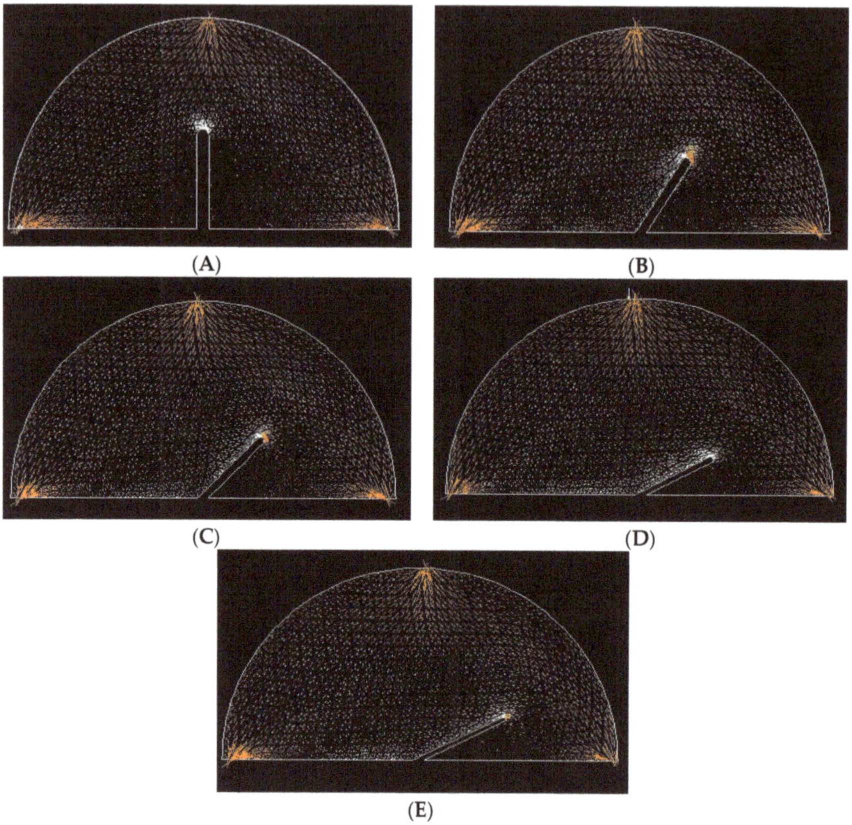

Figure 9. Numerically simulated stress distributions before fracture initiation: (**A**) $\beta = 0°$; (**B**) $\beta = 30°$; (**C**) $\beta = 45°$; (**D**) $\beta = 60°$; (**E**) $\beta = 70°$ (white = tensile stress; orange = compressive stress).

Despite those similar stress distributions for four different β angles, the most significant variations and distinctions in stress distributions were found around the prefabricated chevron notch. Figure 9A shows that for $\beta = 0°$, the tensile stress concentration developed near the chevron notch tip and was distributed around the tip region. When moving away from the tip region, this phenomenon suddenly disappeared along the two faces of the pre-existing notch. Figure 9B–E shows that for $\beta > 30°$, the distributions of tensile stress tended to move toward the upper face of the chevron notch with increasing β. In addition, the tensile stress concentrations were more likely to appear on the left side of the notch tip, while the compressive stress was concentrated on the right side.

In summary, these simulated stress distributions show that except for $\beta = 0°$ resulting in the pure tensile mode (type I), no pure tensile or shear stress distribution was generated, leading to a mixed-mode fracturing process including types I and II. Therefore, both tensile and shear stress were effective for $\beta = 30–70°$ and contributed to the tensile–shearing mixed-mode fracturing propagation in those concrete SCCND specimens. The mechanical SCB tests in previous literature mainly reported and discussed this observation without the assistance of FEM fracturing simulation. Here, investigating the SCB tests with numerical tools can bring more insights at the local scale of the SCCND specimen.

Nevertheless, this series of analyses of the stress distributions only provided the normal stress (tensile and compression) distributions qualitatively. Hence, it remains to analyze the stress distributions on the upper and lower faces of the chevron notch quantitatively in terms of both normal and shear stresses. With this motivation, the "line plot" post-processing option of FRANC2D was used to produce the normal and shear stress distributions around the notch. There were ca. 30 elements around the chevron notch, so the total number of positions was set as 15 for each side.

The quantitative analysis of the normal stress distribution for the chevron notch's upper and lower face is shown in Figure 10A,B, respectively. Figure 10 shows that the wing fracture started at the notch tip because of the fracture formed by tensile stress concentration. For β = 0°, the normal stress was tensile at the notch's upper and lower faces, and the maximum tensile stress (38.5 MPa) was near the tip region. For β = 30°, the normal stress was tensile at the upper face but compressive at the lower face. For β > 30°, the normal stress was slightly tensile when approaching the notch tip on the upper face and then became compressive for both faces near the notch tip. The highest tensile stresses (1–3 MPa) for all β angles were close to the notch tip in the directions perpendicular to the upper face of a fracturing plane. The highest compressive stress (39 MPa) was around the notch's tip in the direction normal to the lower face.

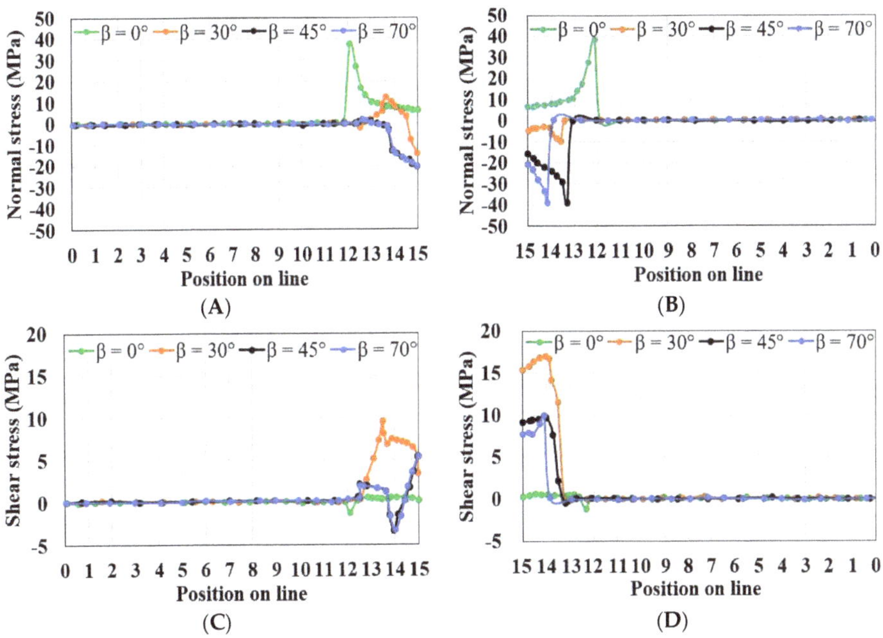

Figure 10. Normal and shear stress distributions around the chevron notch for various values of β: (**A**) normal stress on the upper face; (**B**) normal stress on the lower face; (**C**) shear stress on the upper face; (**D**) shear stress on the lower face (position value equals element number) (the angle β = 60° is not considered due to very close to β = 70°, and there is no significant difference in this regard).

The quantitative analysis of the shear stress distribution for the chevron notch's upper and lower face is shown in Figure 10C,D, respectively. As can be seen, the minimum shear stress was around zero for β = 0° because minimum shear stress contributes to the fracture propagation during such a tensile fracturing process. In contrast, the maximum shear stress reached 15 MPa for β = 30° because the stress mode started to change from the pure tensile mode to the tensile–shearing mixed mode. For $\beta > 30^\circ$, the shear stress direction varied near the notch's tip to form the mixed-mode loading condition around the tip, which was more prone to the shearing mode.

With increasing β, the fracturing mode changed from pure tensile to mixed mode, which was contributed more by shear stress than tensile stress. When β reached 70°, tensile stress contributed insignificantly to fracture initiation, and instead, shear stress was dominant for fracture propagation. This finding agrees with the conclusion from Ayatollahi and Aliha [29] that the pure shearing mode (type II) loading condition can be generated if $\beta \geq 70.53^\circ$ based on the criterion of maximum tangential stress.

3.4. Fracturing in SCCND Simulated by FRANC2D

Fracture-propagation simulations were conducted after the stress-distribution analysis to study how θ varied with β. Note that the automatic re-meshing technique in FRANC2D enabled the simulation of the fracturing process. When a fracture began in the elements near the notch tip, those elements were deleted to allow the fracture to propagate, and a trial mesh was then generated to connect the new fracture to the existing mesh. With due appreciation of this technique, the fracture propagations simulated numerically by FRANC2D for β = 0–70° are compared in Figure 11 with those from the SCB tests.

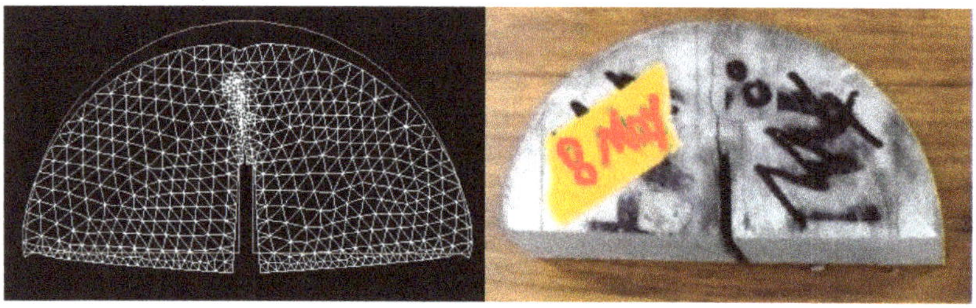

(A)

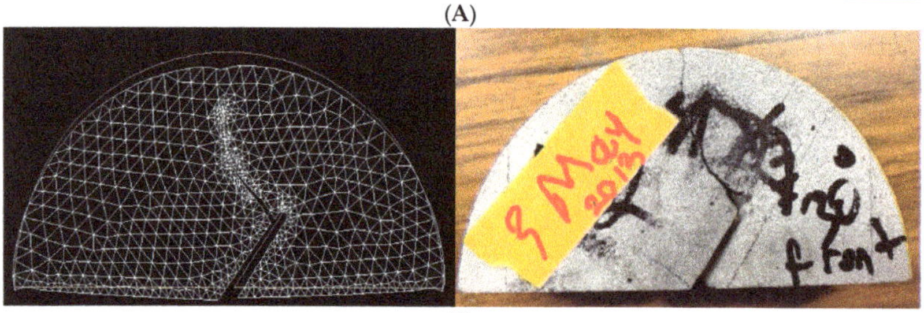

(B)

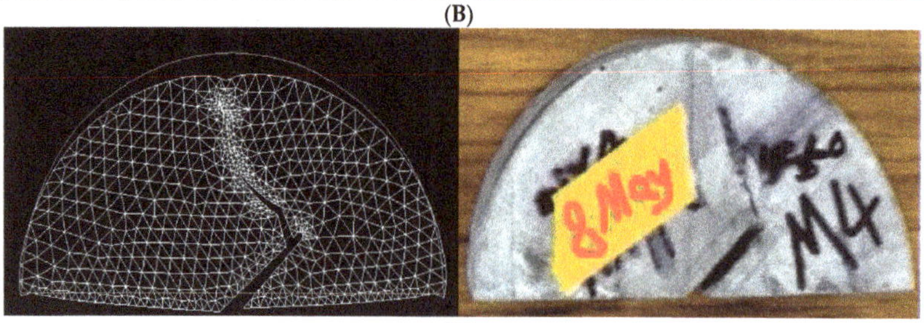

(C)

Figure 11. *Cont.*

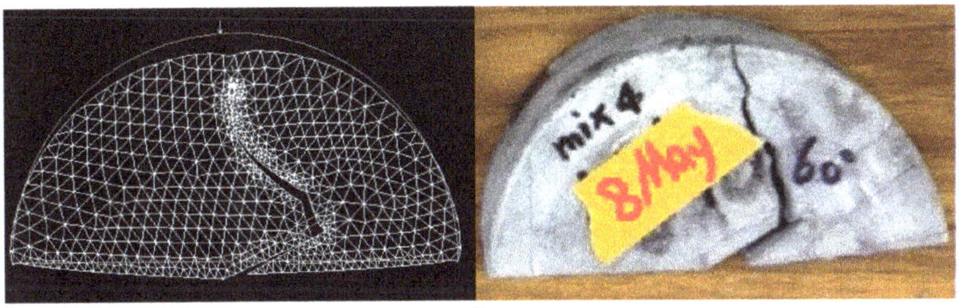

(D)

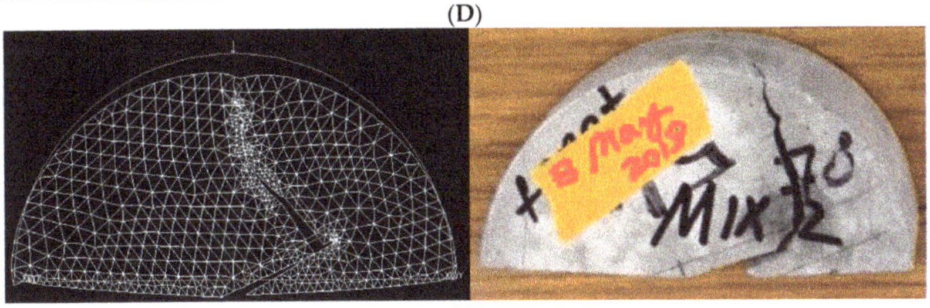

(E)

Figure 11. Fracture propagation simulated numerically by FRANC2D (left) compared with that in SCB tests (right) for various values of β: **(A)** $\beta = 0°$; **(B)** $\beta = 30°$; **(C)** $\beta = 45°$; **(D)** $\beta = 60°$; **(E)** $\beta = 70°$.

Figure 11A–E show good agreements between the simulation and experimental fractures despite minor geometric differences, and all fracturing directions were toward the top loading point. In addition, with increasing β, the wing fractures formed more obviously, and the fracture-initiating point moved from the tip center toward the upper face nearby the tip. This phenomenon agrees with the analysis of stress distributions regarding the tensile stress concentrations moving toward the upper face because the stress-induced fracture always propagates in a direction perpendicular to the local principal tensile stress.

Furthermore, θ depended strongly on β. The θ–β comparison in Table 3 shows that θ increased from 2° to 107° with β increasing from 0° to 70°. The values of θ given by the numerical simulations and the SCB tests agree well for β = 0–30° with a maximum difference of 1°, but they differ more for β = 45–70° with differences of 10–20°. This divergence might have one or more causes, including (i) human operating error, (ii) eccentricity of the top loading point, (iii) local heterogeneity in the concrete specimens (the embedded fine and coarse aggregates), (iv) the mechanical tests being quasi-2D (actually 3D with a thickness B = 30 mm) but the numerical simulations being fully 2D, and (v) meshing issues, but identifying any specific reason was beyond the present scope and could be an academic pursuit for concrete SCB tests in the future.

Table 3. Comparison of simulated and measured values of θ for selected values of β.

β	0°	30°	45°	60°	70°
θ (FRANC2D)	2°	56°	85°	95°	107°
θ (Experiment)	2°	55°	65°	80°	96°

3.5. CMOD Determined by SCB Tests

The CMOD values measured by the COD transducer are plotted in Figure 12 against the load measurement given by the universal loading machine; the load–CMOD sketching was ended before the load reached the relevant value of P_{max}, as given in Table 1. Figure 12 shows that positive CMOD (fracture opening) due to the local principal tensile stress was detected for β = 0–30°. In contrast, negative CMOD (fracture closing) due to the local tensile and shear stresses (mixed mode in types I and II) was detected for β = 30–70°.

The angle β = 30° is a critical inclination angle for the prefabricated chevron notch, at which the positive CMOD turned negative, indicating that the pre-existing notch changed from opening to closing. This opening and closing behavior is also confirmed well by the geometric variation of the notches shown in Figure 11.

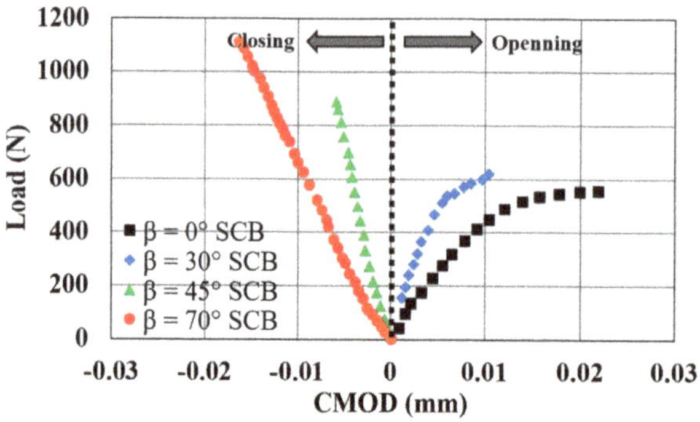

Figure 12. Determination of CMOD in concrete SCCND specimens from SCB tests (the negative value indicates chevron notch closing and the positive value indicates its opening, and the angle β = 60° is not considered due to very close to β = 70°, and there is insignificant difference in this regard).

Also, the load–CMOD curves in Figure 12 can be used in analyzing the fracture-induced plastic deformation of the concrete SCCND specimens. According to Erarslan [14], there is a fracture process zone (FPZ) in front of the chevron notch if there is an apparent plastic deformation in the load–CMOD curve before failure. From Brazilian circular disk (BCD) tests on concrete CCNBD specimens for β = 0–30°, Erarslan [14] concluded that this FPZ occurred only at the notch tip, but it did not occur before the failure point in SCB tests on concrete SCCND specimen for all β angles. Meanwhile, in the present work, Figure 12 shows clearly that this FPZ existed at the notch tip in the SCB tests on the concrete SCCND specimens for β = 0–30°, which is consistent with the BCD tests on concrete CCNBD specimens carried out by Erarslan [14].

3.6. SIF and Fracture Toughness

With the values of $P_{\max}$ given in Table 1, the tensile and shearing SIFs (K_I and K_{II}) were determined as given in Tables 4 and 5 based on the dimensionless SIFs (Y_I and Y_{II}) in Figure 5. To illustrate better how K_I and K_{II} vary with β, the mean values of K_I and K_{II} are plotted in Figure 13A,B, respectively. K_I and K_{II} have minor variations in their STD, demonstrating the reliability and robustness of the SCB tests carried out in this study. Because K_I and K_{II} were calculated based on Y_I and Y_{II} in Figure 5, K_I decreased monotonically from 0.26 MPa·m$^{0.5}$ to 0.07 MPa·m$^{0.5}$ with β increasing from 0° to 70°, while K_{II} increased from zero at 0° to a maximum of 0.10 MPa·m$^{0.5}$ at 45–60° and then decreased to 0.06 MPa·m$^{0.5}$ at 70°. These experimental outcomes agree with the numerical investigation by Erarslan [14], who stated that K_I is always higher than K_{II} for all β angles. Moreover, it is believed that the high STD of 0.127 MPa·m$^{0.5}$ for β = 0° should be due to possible human operating error, the eccentricity of the top loading point, and local heterogeneity in the concrete specimens (the embedded fine and coarse aggregates).

Table 4. Values of tensile SIF (K_I); two SCCND specimens with the same settings were manufactured by splitting each CCNBD specimen equally into first and second half CCNBD specimens.

β	K_I [MPa$\cdot\sqrt{m}$]	Group 1	Group 2	Group 3	Mean	STD
0°	First half CCNBD	0.376	0.116	0.290	0.265	0.127
	Second half CCNBD	0.420	0.077	0.308		
30°	First half CCNBD	0.231	0.181	0.067	0.199	0.063
	Second half CCNBD	0.260	0.233	0.220		
45°	First half CCNBD	–	0.151	0.181	0.177	0.020
	Second half CCNBD	–	0.170	0.206		
60°	First half CCNBD	–	–	0.119	0.131	0.012
	Second half CCNBD	–	–	0.142		
70°	First half CCNBD	–	0.082	–	0.068	0.014
	Second half CCNBD	–	0.054	–		

Table 5. Values of shearing SIF (K_{II}); two SCCND specimens with the same settings were manufactured by splitting each CCNBD specimen equally into first and second half CCNBD specimens.

β	K_{II} [MPa$\cdot\sqrt{m}$]	Group 1	Group 2	Group 3	Mean	STD
0°	First half CCNBD	0.000	0.000	0.000	0.00	0.00
	Second half CCNBD	0.000	0.000	0.000		
30°	First half CCNBD	0.086	0.068	0.025	0.074	0.024
	Second half CCNBD	0.097	0.087	0.082		
45°	First half CCNBD	–	0.084	0.100	0.098	0.011
	Second half CCNBD	–	0.094	0.114		
60°	First half CCNBD	–	–	0.088	0.097	0.009
	Second half CCNBD	–	–	0.105		
70°	First half CCNBD	–	0.072	–	0.060	0.012
	Second half CCNBD	–	0.047	–		

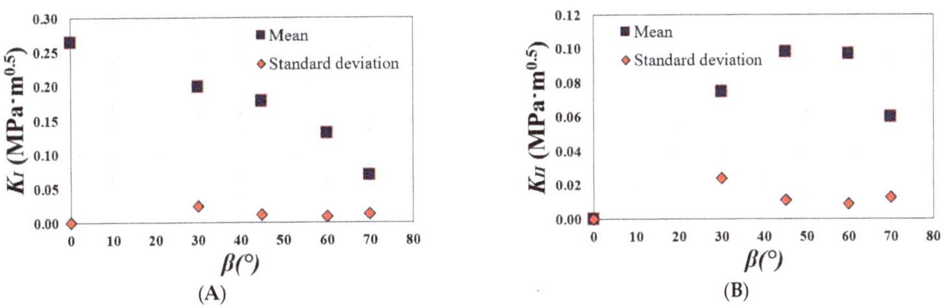

Figure 13. Values of (**A**) K_I and (**B**) K_{II} for various values of β.

The fracture toughness of the tensile mode (K_{Ic}) as calculated by Equations (4) and (5) was 0.375 MPa·m$^{0.5}$ for these concrete SCB tests. However, the determination of K_{IIc} was not straightforward because previous studies have challenged the MTS criterion, which overestimated K_{IIc} for the tensile–shearing mixed-mode loading conditions [14,18,36]. Based on the MTS criterion, K_{IIc} could be given by the maximum fracture toughness of the pure shearing mode, i.e., $K_{II,c} = \left(\sqrt{3}/2\right)K_{Ic}$, if $K_I = 0$ [37]. Regarding the aforementioned stress analysis, it can be seen that diametric compressive loading along the chevron notch would create the mixed-mode (types I and II) fracturing without apparent shearing mode (type II) fracturing, even though the notch would be subjected to compressive–shear loading. However, the MTS and G criteria were derived for a crack under monotonic tension, so earlier criteria for the mixed mode usually fail to predict K_{IIc} obtained experimentally from the SCB tests [29,38,39].

To determine K_{IIc}, the curves of dimensionless K_{II}/K_{Ic} versus K_I/K_{Ic} are shown in Figure 14A, where the MTS, G, GMTS, and extended maximum tangential strain (EMTSN) criteria are compared. As can be seen, the GMTS and EMTSN criteria mostly agree with the experimental results in Figure 14A, unlike the other two criteria, and the EMTSN criterion gives the best prediction. With this finding, it was possible to determine a value for K_{IIc} of ca. 0.075 MPa·m$^{0.5}$ (0.2K_{Ic}). However, although many SCB tests have been conducted previously on other brittle materials, such as rock and glass, there have been few SCB tests on concrete SCCND specimens in a similar experimental setup (sharp tip of notch). A similar value of K_{IIc} = 0.25K_{Ic} in concrete SCB tests was found by Mirsayar, et al. [18], who successfully applied the EMTSN criterion to determine K_{IIc}, although the mortar used and the values of a/R and S/R were not precisely the same as those used in the present work. Nonetheless, by far, the GMTS and EMTSN criteria have been validated as effective mixed-mode fracture criteria to characterize K_{IIc} for marble and concrete [14,18,21,36]. Hence, they are capable of predicting shearing-mode fractures.

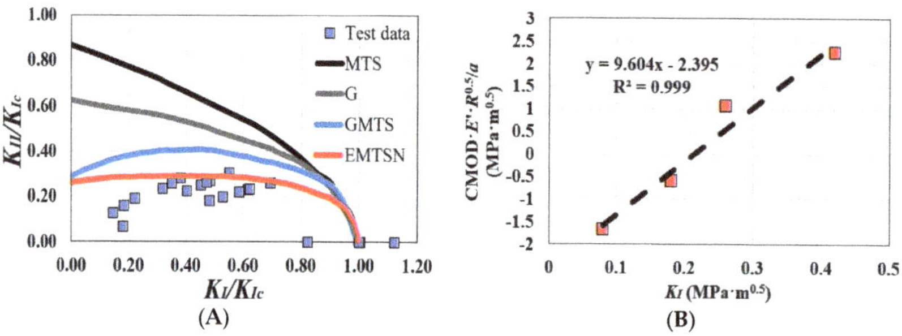

Figure 14. Fracture analysis: (**A**) K_I/K_{Ic} versus K_{II}/K_{Ic} to determine K_{IIc} = 0.3K_{Ic} under GMTS and EMTSN criteria in comparison with conventional criteria, including both MTS and G criteria; (**B**) relationship between CMOD and SIF for concrete SCCND specimens.

However, those criteria are applicable mostly for homogeneous brittle materials such as glass, rock, and concrete without reinforcement. Razmi and Mirsayar [40] and Karimzadeh, et al. [41] used SCB tests to investigate mixed-mode (types I and II) fractures in concrete SCCND specimens reinforced with fibers. Karimzadeh, et al. [41] also concluded that the GMTS criterion outperforms other mixed-mode fracture criteria, but it underestimated K_{IIc} according to their dimensionless K_{II}/K_{Ic}–K_I/K_{Ic} diagram. This underestimation of K_{IIc} might have been due to adding fiber reinforcement to plain concrete, thereby enhancing the mechanical performance, including Young's elastic modulus and fracture toughness.

On the other hand, the relationship between CMOD and K_I was also explored, according to Adamson, et al. [34]. Based on the CMOD–K_I relationship established by Mubaraki and Sallam [35] in Equation (3), Figure 14B shows a perfect linear regression fitting (R^2 = 0.999) between the two variables with the same physical dimensions, with a slope of 9.604 and an intercept of −2.395. This finding also agrees with the linear relationship between CMOD and K_I for pavement material (containing asphalt) reported by Mubaraki and Sallam [35]. Because it was examined for pavement material in the previous study and plain concrete in the present work, this linear relationship might be universal for most brittle materials; nevertheless, more types of brittle materials should be examined by SCB tests in the future to conclude that rigorously.

3.7. Limitations and Prospects

The limitations and prospects of this study can be summarized as follows. First, a few significant STDs were encountered due to possible human operating error, the eccentricity of the top loading point, local heterogeneity in the concrete specimens (the embedded fine and coarse aggregates), and unperfect quasi-2D tests (de facto., 3D tests with a thickness B = 30 mm) compared to the actual 2D plain applicable in analytical analysis and numerical simulations [14]. Second, the uncertainty analysis regarding those issues is worth further pursuing with quantitative analysis, as comprehensively examining various SCB test settings in former studies [31]. Third, as this study narrowed down the research focus merely on mixed-mode fracturing (types I and II) for SCB tests on chevron notch disks having sharp tips instead of a straight-through notch, many other influencing factors, such as the effects of aggregate size [42], cyclic loading [43], another mixed-model fracturing (types I and III) [44], etc., have not been sufficiently considered at the initial stage of this study and should, therefore, be further investigated in the future works. Last, the other numerical method, the discrete element method (DEM) [45], can be additionally involved in future numerical modelling mixed-mode fracturing behavior to explore the dependence of different numerical methods.

4. Summary and Concluding Remarks

In practice, cracked concrete structures sometimes experience mixed-mode loading combining tensile and shearing modes (types I and II). Motivated by this, many SCB tests have been carried out under this mixed-mode loading condition to study concrete fracturing behavior. Many earlier studies were focused on either tensile (type I) loading or this mixed-mode loading condition by using a straight-through notch in pre-cracked concrete specimens. However, despite the success of using a chevron notch with a sharp cracking tip in rock SCB tests, concrete SCB tests have rarely been conducted with such sharp cracking tips. Also, the bluntness of the pre-existing notch tip could undermine the determination of the fracture toughness of the tensile mode (K_{Ic}).

Therefore, the present study investigated mixed-mode (types I and II) fracture propagation in many concrete specimens in SCCNDs both experimentally using SCB tests and numerically using FRANC2D simulations. In total, ten CCNBD specimens with five different chevron notch angles (0–70°) were prefabricated using concrete CCNBD molds, and each circular specimen was split equally into two SCCND specimens with the same inclined angle of the chevron notch (β). Because the present work was focused mainly on mixed-mode fracturing rather than other influential factors, various β angles were manufactured from 0° to 70°, while the other SCB settings remained the same. The SCB tests were carried out using a three-point loading frame, and the CMOD of the notch was measured constantly using a COD transducer. However, because those tests were incapable of detecting the stress distributions in the SCCND specimens, FEM numerical simulations were performed using FRANC2D to replicate them for both comparison and stress-distribution analysis. The following conclusions are drawn based on the results from the SCB tests and FRACN2D simulations.

1. Concrete SCB tests with SCCND specimens with a sharp notch tip are reliable and robust for studying mixed-mode (types I and II) fracture propagation in concrete because of their reliable results and relatively small standard deviations of all testing variables.
2. The maximum failure load ($P_{\max}$) increases with increasing inclination angle (β).
3. The fracture propagation angle (θ) also increases with increasing β, and the phenomenon of wing fractures can be observed.
4. As powerful software for linear elastic fracture mechanics, FRANC2D can successfully simulate SCB tests with the same loading conditions in terms of fracture propagation in SCCND specimens.

5. Despite the fact that the compressive normal stress concentrations in the SCCND specimen are adjacent to the three loading points, most variations of the normal stress distributions occur around the prefabricated chevron notch.
6. With increasing β, the tensile stress concentration around the notch tip moves toward the upper face of the notch, and the compressive stress concentration forms at the notch tip to turn the tensile loading mode into the shearing one.
7. The stress distributions along the upper and lower faces of the chevron notch can be produced by FRANC2D for quantitative analysis, and this stress analysis also verifies the previous findings about stress-distribution variations.
8. The tensile mode (type I) can be generated when β = 0–30° because the CMOD increases, indicating the crack opening under tensile loading. In contrast, the mixed mode (types I and II) becomes more evident for β = 45–70°, with the CMOD decreasing, indicating the crack closing under both tensile and shearing loading conditions.
9. The FPZ can be found for β = 0–30° but not for β = 45–70°, which basically agrees with the CCNBD tests simulated in a previous numerical study.
10. The tensile SIF decreases monotonically with increasing β, whereas the shear SIF increases from zero for β = 0° to a peak value for β = 45–60° and then decreases when β is increased to 70°.
11. Four fracture criteria—MTS, G, GMTS, and EMTSN—were examined against the experimental results. The GMTS and EMTSN criteria outperform the MTS and G ones. In particular, compared with the experimental results, the newly developed EMTSN criterion characterizes most precisely the critical fracture toughness of both tensile and shearing (K_{Ic} and K_{IIc}).
12. There is a linear relationship between CMOD and K_I for concrete SCB tests. However, because it has been examined for both asphalt and concrete, it deserves further investigation for other brittle materials.

In summary, this study investigates concrete SCB tests with a chevron notch (sharp tip) that was initially designed for Brazilian disk tests of rock samples. Based on the testing result, the modified test has been validated as a robust method when applying this type of notch instead of the standard straight-through notch. Numerical fracturing simulations could successfully replicate the fracturing behavior generated by mechanical tests. GMTS and EMTSN criteria have been verified as the two most reliable fracture criteria for concrete material compared to others. In addition, a linear relationship (CMOD-K_I) previously discovered for asphalt SCB tests has been confirmed for concrete SCCND specimens. This work brings more physical insights into mixed-mode fracturing propagation in concrete material.

Author Contributions: Conceptualization, X.L.; Data curation, X.L. and G.Y.; Formal analysis, X.L. and G.Y.; Investigation, X.L.; Methodology, X.L.; Validation, X.L. and G.Y.; Writing—original draft, X.L.; Writing—review & editing, X.L. and G.Y. All authors have read and agreed to the published version of the manuscript.

Funding: This research received no external funding.

Data Availability Statement: The raw data supporting the conclusions of this article will be made available by the authors without undue reservation.

Acknowledgments: This work was supported by the School of Civil Engineering, University of Queensland. In addition, the first author would like to acknowledge the support from Nazife Erarslan. Finally, the first author is also very grateful to Shane Walker and Graham Rule for their assistance at the Material Testing Laboratory, University of Queensland. This work was once presented at the 4th International Congress on Technology Engineering & Science (4th ICONTES) on 5 August 2017 in Kuala Lumpur, Malaysia, and a provisional conference abstract has been collected in the proceedings of the 4th ICONTES (http://procedia.org/cpi/ICONTES-4-2110846).

Conflicts of Interest: The authors declare no conflict of interest.

References

1. Walton, W. Brittle Materials. *Nature* **1958**, *181*, 1577–1579. [CrossRef]
2. Lemaitre, J.; Desmorat, R. Failure of Brittle and Quasi-Brittle Materials. In *Engineering Damage Mechanics*; Springer: Berlin/Heidelberg, Germany, 2005; pp. 321–371.
3. Griffith, A.A., VI. The phenomena of rupture and flow in solids. *Philos. Trans. R. Soc. Lond. Contain. Pap. Math. Phys. Character* **1921**, *221*, 163–198.
4. Whittaker, B.N.; Singh, R.N.; Sun, G. *Rock Fracture Mechanics. Principles, Design and Applications*; Elsevier: Amsterdam, The Netherlands, 1992.
5. Barenblatt, G. Equilibrium cracks formed during brittle fracture rectilinear cracks in plane plates. *J. Appl. Math. Mech.* **1959**, *23*, 1009–1029. [CrossRef]
6. Dugdale, D.S. Yielding of steel sheets containing slits. *J. Mech. Phys. Solids* **1960**, *8*, 100–104. [CrossRef]
7. Lawn, B.R. Fracture of brittle solids. In *Cambridge Solid State Science Series*; Cambridge University Press: Cambridge, UK, 1993.
8. Lim, I.; Johnston, I.; Choi, S. Stress intensity factors for semi-circular specimens under three-point bending. *Eng. Fract. Mech.* **1993**, *44*, 363–382. [CrossRef]
9. Shah, S.P.; Swartz, S.E.; Ouyang, C. *Fracture Mechanics of Concrete: Applications of Fracture Mechanics to Concrete, Rock and Other Quasi-Brittle Materials*; John Wiley & Sons: Hoboken, NJ, USA, 1995.
10. Bocca, P.; Carpinteri, A.; Valente, S. Mixed mode fracture of concrete. *Int. J. Solids Struct.* **1991**, *27*, 1139–1153. [CrossRef]
11. Sih, G.C. Strain-energy-density factor applied to mixed mode crack problems. *Int. J. Fract.* **1974**, *10*, 305–321. [CrossRef]
12. Shen, B.; Stephansson, O. Cyclic loading characteristics of joints and rock bridges in a jointed rock specimen. In Proceedings of the International Symposium on Rock Joints, Loen, Norway, 4–6 June 1990; pp. 725–729.
13. Lim, W.K.; Lee, C.S. Evaluation of stress intensity factors for a crack normal to bimaterial interface using isoparametric finite elements. *Eng. Fract. Mech.* **1995**, *52*, 65–70.
14. Erarslan, N. Analysing mixed mode (I–II) fracturing of concrete discs including chevron and straight-through notch cracks. *Int. J. Solids Struct.* **2019**, *167*, 79–92. [CrossRef]
15. Ayatollahi, M.; Aliha, M.; Saghafi, H. An improved semi-circular bend specimen for investigating mixed mode brittle fracture. *Eng. Fract. Mech.* **2011**, *78*, 110–123. [CrossRef]
16. Shi, Z. *Crack Analysis in Structural Concrete: Theory and Applications*; Butterworth-Heinemann: Oxford, UK, 2009.
17. Bažant, Z.P.; Oh, B.H. Crack band theory for fracture of concrete. *Matériaux Constr.* **1983**, *16*, 155–177. [CrossRef]
18. Mirsayar, M.; Razmi, A.; Berto, F. Tangential strain-based criteria for mixed-mode I/II fracture toughness of cement concrete. *Fatigue Fract. Eng. Mater. Struct.* **2018**, *41*, 129–137. [CrossRef]
19. Aliha, M.; Mahdavi, E.; Ayatollahi, M. Statistical analysis of rock fracture toughness data obtained from different chevron notched and straight cracked mode I specimens. *Rock Mech. Rock Eng.* **2018**, *51*, 2095–2114. [CrossRef]
20. Xu, Y.; Dai, F.; Zhao, T.; Xu, N.-W.; Liu, Y. Fracture toughness determination of cracked chevron notched Brazilian disc rock specimen via Griffith energy criterion incorporating realistic fracture profiles. *Rock Mech. Rock Eng.* **2016**, *49*, 3083–3093. [CrossRef]
21. Luo, L.; Erarslan, N.; Mu, Z. Mixed-Mode Fracturing of Concrete—Numerical and Experimental Study using SCCD Specimens. In *Recent Advances in Material, Analysis, Monitoring, and Evaluation in Foundation and Bridge Engineering*; ASCE Library: Reston, VA, USA, 2014; pp. 25–37.
22. Kuruppu, M.D.; Obara, Y.; Ayatollahi, M.R.; Chong, K.; Funatsu, T. ISRM-suggested method for determining the mode I static fracture toughness using semi-circular bend specimen. *Rock Mech. Rock Eng.* **2014**, *47*, 267–274. [CrossRef]
23. Wawrzynek, P.A.; Ingraffea, A. Interactive finite element analysis of fracture processes: An integrated approach. *Theor. Appl. Fract. Mech.* **1987**, *8*, 137–150. [CrossRef]
24. Henshell, R.; Shaw, K. Crack tip finite elements are unnecessary. *Int. J. Numer. Methods Eng.* **1975**, *9*, 495–507. [CrossRef]
25. Gómez, F.; Elices, M. A fracture criterion for blunted V-notched samples. *Int. J. Fract.* **2004**, *127*, 239–264. [CrossRef]
26. Chen, D.; Ozaki, S. Investigation of failure criteria for a sharp notch. *Int. J. Fract.* **2008**, *152*, 63–74. [CrossRef]
27. Erarslan, N.; Williams, D. Mixed-mode fracturing of rocks under static and cyclic loading. *Rock Mech. Rock Eng.* **2013**, *46*, 1035–1052. [CrossRef]
28. Ayatollahi, M.R.; Aliha, M.R.M.; Hassani, M.M. Mixed mode brittle fracture in PMMA—An experimental study using SCB specimens. *Mater. Sci. Eng. A* **2006**, *417*, 348–356. [CrossRef]
29. Ayatollahi, M.R.; Aliha, M.R.M. On determination of mode II fracture toughness using semi-circular bend specimen. *Int. J. Solids Struct.* **2006**, *43*, 5217–5227. [CrossRef]
30. Kuruppu, M.D.; Chong, K.P. Fracture toughness testing of brittle materials using semi-circular bend (SCB) specimen. *Eng. Fract. Mech.* **2012**, *91*, 133–150. [CrossRef]
31. Ayatollahi, M.R.; Aliha, M.R.M. Wide range data for crack tip parameters in two disc-type specimens under mixed mode loading. *Comput. Mater. Sci.* **2007**, *38*, 660–670. [CrossRef]
32. Chang, S.-H.; Lee, C.-I.; Jeon, S. Measurement of rock fracture toughness under modes I and II and mixed-mode conditions by using disc-type specimens. *Eng. Geol.* **2002**, *66*, 79–97. [CrossRef]
33. Lanaro, F.; Sato, T.; Stephansson, O. Microcrack modelling of Brazilian tensile tests with the boundary element method. *Int. J. Rock Mech. Min. Sci.* **2009**, *46*, 450–461. [CrossRef]

34. Adamson, R.; Dempsey, J.; Mulmule, S. Fracture analysis of semi-circular and semi-circular-bend geometries. *Int. J. Fract.* **1996**, *77*, 213–222. [CrossRef]
35. Mubaraki, M.; Sallam, H. Reliability study on fracture and fatigue behavior of pavement materials using SCB specimen. *Int. J. Pavement Eng.* **2020**, *21*, 1563–1575. [CrossRef]
36. Aliha, M.; Ayatollahi, M.R. Mixed mode I/II brittle fracture evaluation of marble using SCB specimen. *Procedia Eng.* **2011**, *10*, 311–318. [CrossRef]
37. Erdogan, F.; Sih, G. On the crack extension in plates under plane loading and transverse shear. *J. Basic Eng.* **1963**, *85*, 519–525. [CrossRef]
38. Backers, T. *Fracture Toughness Determination and Micromechanics of Rock under Mode I and Mode II Loading*; GeoForschungsZentrum: Potsdam, Germany, 2005.
39. Rao, Q.; Sun, Z.; Stephansson, O.; Li, C.; Stillborg, B. Shear fracture (Mode II) of brittle rock. *Int. J. Rock Mech. Min. Sci.* **2003**, *40*, 355–375. [CrossRef]
40. Razmi, A.; Mirsayar, M. On the mixed mode I/II fracture properties of jute fiber-reinforced concrete. *Constr. Build. Mater.* **2017**, *148*, 512–520. [CrossRef]
41. Karimzadeh, H.; Razmi, A.; Imaninasab, R.; Esminejad, A. The influence of natural and synthetic fibers on mixed mode I/II fracture behavior of cement concrete materials. *Can. J. Civ. Eng.* **2019**, *46*, 1081–1089. [CrossRef]
42. Golewski, G.L. Effect of Coarse Aggregate Grading on Mechanical Parameters and Fracture Toughness of Limestone Concrete. *Infrastructures* **2023**, *8*, 117. [CrossRef]
43. Alrayes, O.; Könke, C.; Hamdia, K.M. A Numerical Study of Crack Mixed Mode Model in Concrete Material Subjected to Cyclic Loading. *Materials* **2023**, *16*, 1916. [CrossRef] [PubMed]
44. Sreenath, S.; Mohan, K.S.R.; Murali, G. Fracture Toughness of Reactive Powder Fibrous Concrete Composites under Pure and Mixed Modes (I/III). *Buildings* **2022**, *12*, 599. [CrossRef]
45. Zhu, R.; Gao, H.; Zhan, Y.; Wu, Z.-X. Construction of Discrete Element Constitutive Relationship and Simulation of Fracture Performance of Quasi-Brittle Materials. *Materials* **2022**, *15*, 1964. [CrossRef]

buildings

MDPI

Article

Plastic Zone Radius Criteria for Crack Propagation Angle Evaluated with Experimentally Obtained Displacement Fields

Jorge Guillermo Díaz-Rodríguez [1,*], Alberto David Pertúz-Comas [2], Oscar Rodolfo Bohórquez-Becerra [3], Arthur Martins Barbosa Braga [4] and Darío Prada-Parra [4]

1 Escuela de Ingeniería y Ciencias, Tecnológico de Monterrey, Guadalajara 45201, Mexico
2 GIC, Grupo de Investigaciones en Corrosión, Escuela de Ingeniería Mecánica, Universidad Industrial de Santander, Bucaramanga 680002, Colombia; apertuzc@uis.edu.co
3 GIEMA, Grupo de Investigación en Energía y Medio Ambiente, Escuela de Ingeniería Mecánica, Universidad Industrial de Santander, Bucaramanga 680002, Colombia; obohorbe@correo.uis.edu.co
4 Department of Mechanical Engineering, Pontifical Catholic University of Rio de Janeiro, Gávea, Rio de Janeiro 22451-900, Brazil; abraga@puc-rio.br (A.M.B.B.); daprada@puc-rio.br (D.P.-P.)
* Correspondence: jorgegdiaz@tec.mx

Abstract: The monitoring and maintenance of cracked structures are generally carried out using structural integrity assessments. The plastic zone (PZ) crack path (CP) criteria state that a crack grows in a specific direction when the radius of the plastic zone ahead of the crack tip reaches a minimum value. The PZ can be evaluated using stress intensity factors (SIFs). The SIFs under mixed-mode loading were extracted from the literature from three samples: two single edge notch tension (SENT) samples (E = 2.5 GPa, v = 0.38) made from polycarbonate and one modified compact test (C(T)) sample made from low-carbon steel (E = 200 GPa, v = 0.3). In addition, the CP angle was evaluated for the W and R criteria with experimental data, which included non-linear effects such as fatigue-induced plasticity, crack roughness, and debris. It was found that both can predict the CP for lateral cracks in both tested materials and monotonic and cyclic load when the mode mixity does not change considerably from one crack length to the next or goes beyond 0.2. Moreover, the R criterion exhibited an error as high as 1.7%, whereas the W criterion showed a 6% error on the last crack length for the low-carbon steel sample under cyclic load, which had a 100% increase in mode mixity. Finally, the applicability of LEFM was checked, while the CP was sought by finding the size of the PZ.

Keywords: LEFM; crack kinking; crack path; plastic zone

1. Introduction

In the maintenance of cracked components and structures, the crack front advance rate needs to be established and monitored [1–3]. However, in the case of mixed-mode loading, factors such as plasticity [4], crack interlocking, crack rugosity [5], trapped debris [6], and crack tip vicinity to defects [7], among others, may affect the crack growth rate and crack path (CP). Thus, CP prediction is not entirely understood [1,2,4,5,8–10]. According to [11], the more kinked the crack is, the more crack growth might be retarded. Nevertheless, for crack kinking and equivalent SIFs, there are already several postulated criteria [1,12], mainly based on linear elastic fracture mechanics (LEFM), with the maximum tangential stress criteria being the most used for mixed-mode loading [10]. However, the applicability of LEFM must be checked [13,14] before a suitable model can be applied. Using a model that checks the applicability and predicts CP simplifies those two tasks.

Overall, it can be said that all CP prediction models state a hypothesis for a crack to change direction by a small $\Delta\theta$ from its original path. Such an increment is generally created when an equivalent SIF (Keq) exceeds material toughness (K_{Ic}) [10,13] in the LEFM approach. Wasiluk and Golos [8] proposed a model based on a dimensionless plastic zone radius, P_{zr}. Ren et al. [15] proposed a model based on the plastic zone radius and evaluated it for plane stress and plane strain conditions. Both models were devised for a central crack. Comprehensive reviews of CP models can be found in [1,12].

Although the CP models described here use SIFs (monotonic load), in the literature [12,16,17], a swap between monotonic SIF (K) and fatigue SIF range (ΔK) for proportional load was found to be acceptable. This means that a crack grows under fatigue loading if ΔKeq exceeds the fatigue threshold (ΔKth), or the crack grows under static loading when the equivalent SIF (Keq) exceeds the fracture toughness (Kc), as has been postulated before [16]. Finally, we can say that SENT samples have gained popularity in testing pipelines [18] and welds or offer less restriction [6,19], and C(T) samples offer a larger space to observe CP evolution [20–22]. Lately, some researchers have devoted efforts to evaluating CP in complex problems using artificial intelligence, such as heterogeneous materials [7], metamaterials [23], or composites [24]. However, the computational cost might not be worth it for isotropic materials where parametric and less computationally costly models are sufficient for prediction.

Although both evaluated models were proposed for monotonic central cracks in infinite plates, this paper evaluates them on lateral cracks and, at the same time, assesses the applicability of LEFM in fatigue and monotonic loading. Furthermore, the models are evaluated with experimental data that include non-linear effects such as fatigue-induced plasticity (which induces hardening ahead of the crack tip, which may produce crack retardation), crack roughness, crack interlocking (which might hinder the full development of mode II), and trapped debris (which keep the crack open, thus accelerating mode I propagation). Our literature review revealed that the combination of such non-linearities cannot yet be numerically reproduced. Therefore, the first set of experimental data is for proportional fatigue load [25], whereas the second set is for proportional monotonic load [26], both under different mode mixity ratios (K_{II}/K_I). The first data set was obtained using the digital image correlation (DIC) technique, whereas the second set was obtained through finite element modeling (FEM) and validated with photoelasticity. The performance of both CP models is shown along the crack length.

The rest of the paper is organized as follows: Section 2 describes the basis for the CP used; Section 3 explains the experimental details of where the data were obtained from; Section 4 groups the CP prediction and discusses the results attributable to the appropriate phenomena; and Section 5 draws the conclusions and presents the findings.

2. Criteria for the Crack Path

The stress field in a cracked body under an applied force F is schematically represented in Figure 1, where σ and τ are the normal and shear stress, respectively, r and θ are the radial and tangential directions, respectively, and a is the crack length.

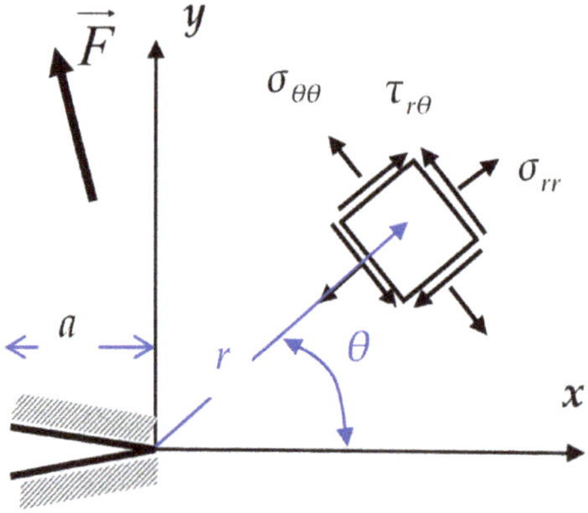

Figure 1. Schematics and notation of acting stress in a cracked body.

In 2000, Wasiluk and Golos proposed the W criterion [8]. It states that a central crack under mixed-mode loading grows in the direction where the radius from the crack tip to the boundary of the plastic zone is the shortest [8]. They started from the radius of the plastic zone, r_{pz}, proposed by Pook [27] for mode I and II loading and compared the stress field with the von Misses yield criteria for a plate with a centrally inclined crack, as shown in Equation (1).

$$r_{pz} = \frac{1}{2\pi\sigma_y{}^2}\left[K_I{}^2\cos^2\left(\frac{\theta}{2}\right)\left(1+3\sin^2\left(\frac{\theta}{2}\right)\right)+K_IK_{II}\sin\theta(3\cos\theta-1)+K_{II}{}^2\left(3+\sin^2\left(\frac{\theta}{2}\right)\left(1-9\sin^2\left(\frac{\theta}{2}\right)\right)\right)\right] \tag{1}$$

where σ_y is the yield stress and K_I and K_{II} are the SIFs in modes I and II, respectively. Then, W is the factor of dividing r_{pz} over half the crack length, and using $K_I = \sigma\sin2\theta\sqrt{\pi a}$ and $K_{II} = \sigma\sin\theta\cos\theta\sqrt{\pi a}$, we obtained Equation (2).

$$W = \frac{\sigma^2}{2\sigma_y{}^2}\left[\begin{array}{l}\sin^4(\theta_W)\cos^2\left(\frac{\theta}{2}\right)\left(1+3\sin^2\left(\frac{\theta}{2}\right)\right)+\\ +\sin^4(\theta_W)\sin(\theta)(3\cos\theta-1)+\sin^2(\theta_W)\cos^2(\theta_W)\left(3+\sin^2\left(\frac{\theta}{2}\right)\left(1-9\sin^2\left(\frac{\theta}{2}\right)\right)\right)\end{array}\right] \tag{2}$$

Thus, taking the first and second derivative of W, Equation (2) with respect to θ, as seen in Equation (3), gives a W_{min} at which the direction of crack growth, θ_W, will happen.

$$\left.\frac{\partial W_{(\theta,\theta*,\sigma)}}{\partial\theta}\right|_{\theta=\theta*} = 0 \text{ and } \left.\frac{\partial^2 W_{(\theta,\theta*,\sigma)}}{\partial\theta^2}\right|_{\theta=\theta*} > 0 \tag{3}$$

Then, if the applied load is cyclic, SIF, K, becomes the SIF range, ΔK, and the fracture toughness, Kc, becomes the fatigue threshold, ΔKth, as proposed and tested in [17], so Equation (1) can be rewritten as Equation (4).

$$r_{pz-W} = \frac{1}{2\pi\sigma_y{}^2}\left[\begin{array}{l}\Delta K_I{}^2\cos^2\left(\frac{\theta}{2}\right)\left(1+3\sin^2\left(\frac{\theta}{2}\right)\right)+\Delta K_I\Delta K_{II}\sin\theta(3\cos\theta-1)+\\ +\Delta K_{II}{}^2\left(3+\sin^2\left(\frac{\theta}{2}\right)\left(1-9\sin^2\left(\frac{\theta}{2}\right)\right)\right)\end{array}\right] \tag{4}$$

On the other hand, in 2014, Ren [15] proposed a model based on the PZ radius and evaluated it for plane stress and plane strain conditions for a centrally inclined crack. The radius of the plastic zone is shown in Equation (5).

$$r_{pz} = \frac{1}{2\pi\sigma_y{}^2}\left[\begin{array}{l}K_I{}^2\frac{2(1-2\nu)^2}{3}\left[(1+\cos\theta)+\sin^2\theta\right]+2K_IK_{II}\left(\sin2\theta-\frac{2(1-2\nu)^2}{3}\right)+\\ +K_{II}{}^2\left(1+\frac{2(1-2\nu)^2}{3}\right)(1-\cos\theta)(3+\cos^2\theta)\end{array}\right] \tag{5}$$

where ν is Poisson's modulus. Moreover, assuming the interchangeability of K and ΔK and Kc with ΔKth, Equation (5) for monotonic loading can be rewritten as Equation (6) for fatigue loading.

$$r_{pz} = \frac{1}{2\pi\sigma_y{}^2}\left[\begin{array}{l}\Delta K_I{}^2\frac{2(1-2\nu)^2}{3}\left[(1+\cos\theta)+\sin^2\theta\right]+2\Delta K_I\Delta K_{II}\left(\sin2\theta-\frac{2(1-2\nu)^2}{3}\right)+\\ +\Delta K_{II}{}^2\left(1+\frac{2(1-2\nu)^2}{3}\right)(1-\cos\theta)(3+\cos^2\theta)\end{array}\right] \tag{6}$$

where ΔK is the SIF range, and ΔKth is the fatigue threshold. It is noted how the PZ can give an insight into crack initiation at the shortest radius. Furthermore, an analysis of Equations (4) and (6) gives the angles at which the minimum radius occurs. This is 0° for mode I and 82.5° for pure mode II in the R criterion and 77° in the W criterion. Furthermore, Figure 2 shows how the PZ changes for different mode mixity (K_{II}/K_I) ratios for both the W and R criteria, with the latter plotted for plain stress. The PZ is plotted for different mode mixity ratios, from pure mode I, $K_{II}/K_I = 0$, until pure mode II, $K_{II}/K_I = \infty$, but keeping Keq constant, $K_{eq} = \sqrt{(K_I^2 + K_{II}^2)} = 1$. Both K_I and K_{II} were always kept positive. A negative K_I does not make physical sense as it requires the crack lips to overlap, whereas a negative K_{II} implies the relative displacements of two opposite-to-crack points. It can be seen how the PZ changes the applied load's direction.

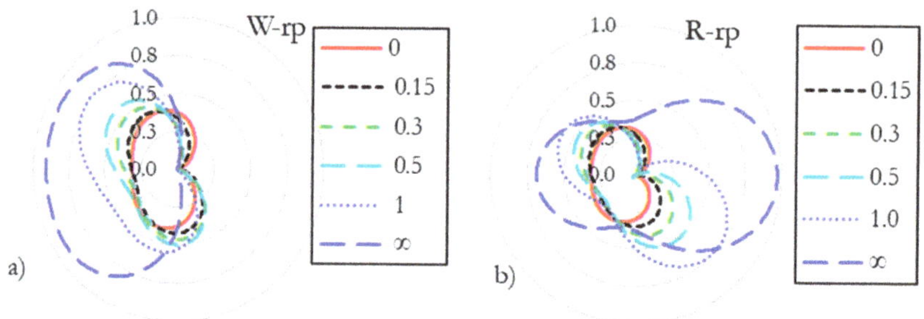

Figure 2. Dimensionless PZ for different mode mixity ratios (K_{II}/K_I). (**a**) W criterion; (**b**) R criterion.

One can see in Figure 2 that the crack orientation of the PZ changes from perpendicular to the tensile stress, mode I, to parallel to the shear stress, mode II. Therefore, a parameter, M [9,28], is used to tell whether the crack growth is dominated by the tensile mode or the shear mode, as shown in Equation (7).

$$M = \frac{2}{\pi} \tan^{-1}\left(\frac{K_{II}}{K_I}\right) \quad (7)$$

3. Materials and Methods

The data used to evaluate the two CP models were obtained from two samples subjected to proportional loading tests. The first was an 8.7 mm thick modified compact test C(T) specimen made out of low-carbon steel (E = 200 GPa, v = 0.3) with a drilled hole in front of the CP to modify the stress field [25], as seen in Figure 3a, with the testing setup shown in Figure 3b. Miranda tested a similar sample [29], with the results computationally verified later on [22]. The experimentally observed CP for this sample and the six measured points are shown in Figure 3a. The hole modified the stress field, which curved the crack and induced opening mode II over the applied mode I load. Moreover, because the load inversion ratio was 0.1, the SIF range (ΔK) was calculated as Kmax–Kmin from DIC displacement measurements and verified through FEM [30]. Table 1 shows this sample's retrieved crack length, CP angle, SIF ranges, and mode mixity for the six measured points. Finally, the fatigue loading rate for the modified C(T) sample was applied at 10 Hz, and for about every 1 mm of crack growth, the load frequency was lowered to 0.1 Hz to allow for image recording for DIC analysis.

Because the applied load on this sample was cyclic, Equations (4) and (6) were used to establish the sample's PZ radius and shape. Extensive details about this test and specifics on how the SIFs were calculated can be found in [25].

As stated before, the displacement fields for the modified C(T) were measured using the DIC technique. It provides independent displacement fields using non-contact optical measurements. Once displacements are obtained, one can compute strains, and by using a constitutive law, stresses can be calculated. In-depth details for DIC, including its capabilities, limitations, and implementation, can be found in [30].

Figure 3. Schematics of the tests. (**a**) Measured points for the holed CT specimen; (**b**) testing setup for the holed CT sample; (**c**) SENT with the inclined crack. (**a**,**b**) from [25].

Table 1. Experimental parameters for the modified C(T) at different crack lengths, from [25].

Point	a [mm]	θ*	ΔK_I, MPa$\sqrt{m}$	ΔK_{II}, MPa$\sqrt{m}$	$\Delta K_{II}/\Delta K_I$
0	2.1	0	13.12	0.46	0.04
a	4.1	0	17.78	0.47	0.03
b	6.3	−5	18.14	0.59	0.03
c	8.2	−5	19.67	1.17	0.06
d	10.2	−7	22.00	1.55	0.07
e	11.9	−24	26.85	3.55	0.13

The second type of samples was two 3 mm thick polycarbonate (E = 2.5 GPa, v = 0.38) SENT samples [26], see the schematics in Figure 3c, subjected to simple remote tension with a 22.5°crack and another with a 45° inclined pre-existing crack producing mixed-mode (I and II) loading conditions. The retrieved data included SIF modes I and II, obtained through numerical simulation and validated with photoelasticity, shown in Table 2, where a/W is the dimensionless residual ligament. Because the applied loading on these samples was monotonic, Equations (1) and (5) were used to establish the PZ radius and shape.

Table 2. Experimental parameters for the thin strips from [26].

a/W	$K_I/(\sigma\sqrt{\pi a})$	$K_{II}/(\sigma\sqrt{\pi a})$	K_{II}/K_I
	45°		
0.3	0.83	0.48	0.58
0.4	0.96	0.50	0.52
0.5	1.22	0.57	0.47
0.6	1.42	0.65	0.46
	22.5°		
0.3	1.48	0.35	0.24
0.4	1.80	0.41	0.23
0.5	2.28	0.54	0.24
0.6	2.99	0.68	0.23

Finally, the performance of the CP models was evaluated with the expression proposed in Equation (8), as recently proposed [31].

$$err = \frac{\theta_{\text{exp}} - \theta*}{\pi} \quad (8)$$

where θexp is the experimental CP angle and θ* is the angle predicted by either model.

4. Results and Discussion

4.1. CP Prediction for Modified C(T)

Figure 4 compares the experimentally observed CP for the W and R criteria predicted angle for the six measured points in the modified C(T) sample. One can see that the prediction is very close for short cracks as the mode mixity ratio, $\Delta K_{II}/\Delta K_I$, is low. However, at the last measurement before unstable crack growth, the $\Delta K_{II}/\Delta K_I$ rises to 0.13, and the predicted angle gives a 6% error for the W criterion. In contrast, the R criterion shows a 1.7% error. The incremental change in mode mixity is attributed to the machined hole, which caused an asymmetry to the stress field, producing larger stresses toward the hole that ultimately caused the crack to kink towards it. As the crack grew, the stress field became even more asymmetric, continuously increasing K_{II}, kinking the crack more pronouncedly every loading cycle until it finally met the hole.

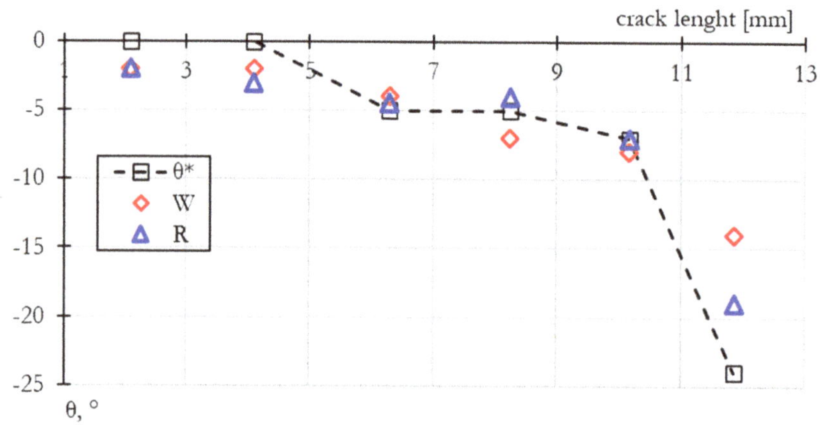

Figure 4. Evaluation of crack kinking models for the modified C(T) sample.

The W criteria made a close prediction for most of the crack lengths. Figure 5 shows the comparative size and shape of the PZ, Equation (4) W criterion, and the Equation (5) R criterion, for six different crack lengths in this specimen. It can be seen how the PZ orientation turns as the crack length increases and the mode mixity changes. This turning is attributed to the increase in ΔK_{II}, which makes the crack deviate from the applied mode. I load Moreover, it can be observed that the size of the largest PZ, about 2.54 mm for the W criterion and about 2.9 mm for the R criterion, in their largest axis, which are below the thickness of the specimen, 8.7 mm. Thus, LEFM adequately describes the stress field. For this sample, the W criterion showed larger PZ for the first five crack sizes, whereas the R criterion retrieved a more extensive zone for the last one, where K_{II} reaches 3.55 MPa$\sqrt{m}$ and the mode mixity ratio, K_{II}/K_I, is 0.13.

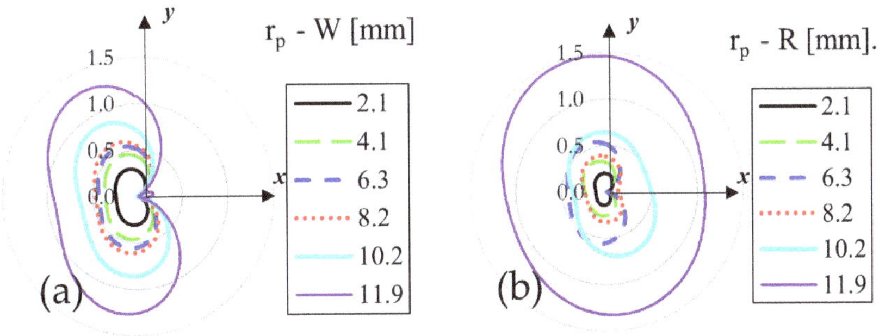

Figure 5. Size and shape or plastic zone for the modified C(T) specimen at different crack lengths: (**a**) from the W criterion and (**b**) from the R criterion.

4.2. CP Prediction for SENT Samples

On the other hand, the performance of the CP models for both SENT specimens for dimensionless crack length is shown in Figure 6. The models were evaluated with the retrieved K_I and K_{II}. Unlike the C(T) sample, the applied load is monotonic, so there is no need to assume that K is ΔK and K_c is ΔK_{th}. The SENT with a 45° crack showed a K_{II}/K_I ratio from 0.46 at the longest to 0.58 for the shortest a/W. Polycarbonate is a material prone to shear-dominated failure, and the 45° crack induces a high K_{II}/K_I ratio. Therefore, the W criterion is not expected to perform well under these conditions, so it predicts a quasi-constant 29°, whereas the R criterion predicts a closer CP between 43 and 47°.

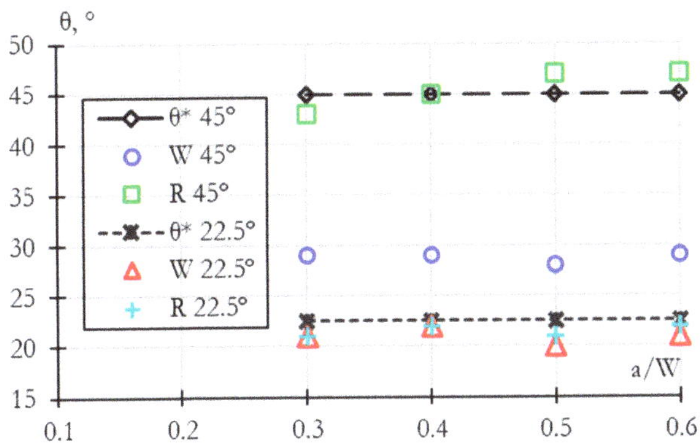

Figure 6. Evaluation of crack kinking models for SENT samples.

Conversely, the SENT with a 22.5° crack showed a quasi-constant K_{II}/K_I ratio of 0.23. Because of this constant K_{II}/K_I ratio, the predicted CP angle is also quasi-constant, about 21° degrees for the W criterion, very close to the experimentally observed angle of 22.5°. Conversely, the R criterion predicts a quasi-constant angle of 17°.

The shape and size of the plastic zone are shown in Figure 7 for the SENT sample with an initial crack of 45° for the four different reported crack lengths. It can be seen how the orientation of the plastic zone stays pretty much constant as the crack length increases for the W and R criteria. This situation is attributed to the steady K_{II}/K_I ratio, as seen in Table 2, making the crack grow straight from its original path, even though only mode I is applied. Moreover, it can be observed that the size of the largest plastic zone on its largest axis is about 0.7 mm for the W criterion and 0.58 mm for the R criterion. In both cases, this size is below the thickness of the specimen, 3.0 mm, so LEFM can be used to describe the stress field satisfactorily. Finally, it is noted that the W criterion predicts a slightly larger PZ than the R criterion.

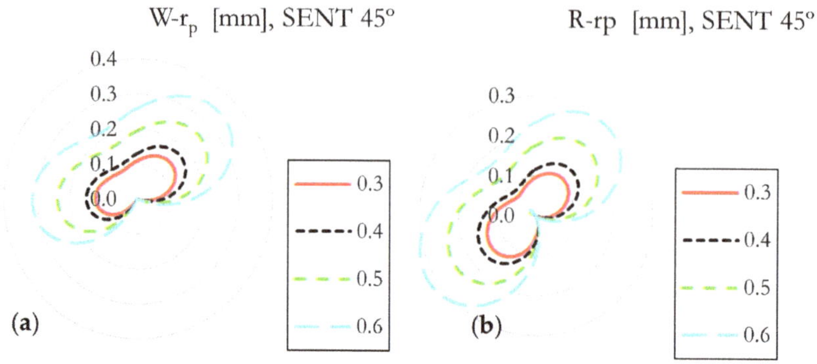

Figure 7. Size and shape of the plastic zone for the SENT sample 45°: (**a**) from the W criterion, (**b**) from the R criterion.

Figure 8 shows the PZ size for the SENT sample with an initial crack of 22.5° for the four different reported crack lengths. It can be seen how the orientation of the plastic zone stays pretty much constant as the crack length increases. This situation is attributed to the steady K_{II}/K_I ratio, making the crack grow straight from its original inclined path, even though there is an applied load in mode I. Moreover, it was observed that the size of the largest PZ, about 1.8 mm for the W criterion and about 1.54 mm for the R criterion in its largest axis, is below the thickness of the specimen, 3.0 mm, so LEFM can be used to properly describe the stress field. In this sample, it was also observed that the W criterion predicts a slightly larger PZ than the R criterion.

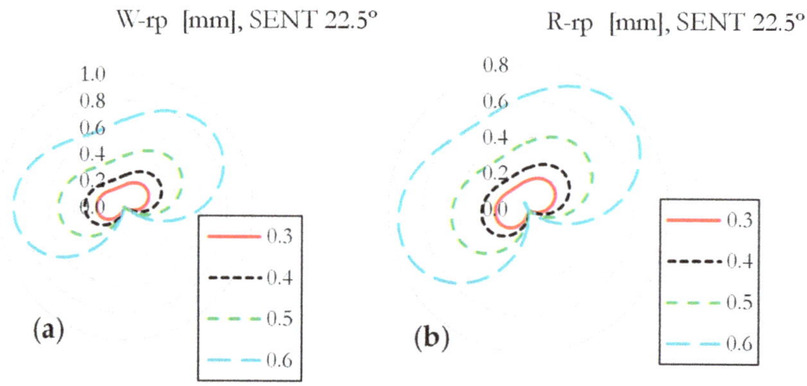

Figure 8. Size and shape of the plastic zone for the SENT sample 22.5°: (**a**) from the W criterion and (**b**) from the R criterion.

4.3. Discussion

Highsmith [10] stated that a crack under mixed mode will grow when the equivalent SIF exceeds the material toughness, Kc, or the equivalent SIF range exceeds the material fatigue threshold, K_{th}. In both cases, the crack grew under either monotonic or fatigue loading. Therefore, the Keq was larger than the Kc for the monotonic load, and the ΔKeq was larger than the Kth for the cycling load. However, because more than one loading mode was present, the CP must be established to describe crack growth completely [31].

On the other hand, Vormwald et al. [9] argued that the mode mixity ratio [28] might be the parameter that can be used to tell when a crack growth scenario can change from tensile to shear-dominating load. It could make a crack turn when it reaches 0.5. Figure 9 shows the variation in the CP angle with mode mixity for both samples. For the SENT samples, neither the load nor the mode mixity changed much with crack length. For the 22.5° SENT sample, the K_{II}/K_I ratio kept a 0.24 constant value, whereas for the 45° sample, the K_{II}/K_I ratio went from 0.56 to 0.46, as seen in Figure 6. Vormwald et al. [9] argued that when transitioning from tensile to entirely dominated shear crack growth, such mode mixity parameters might not be able to tell the CP. Therefore, the CP for the SENT samples was not expected to deviate from its initial angle. This can be seen in Figure 9a for the three samples and Figure 9b for a close-up of the SENT 22.5° sample.

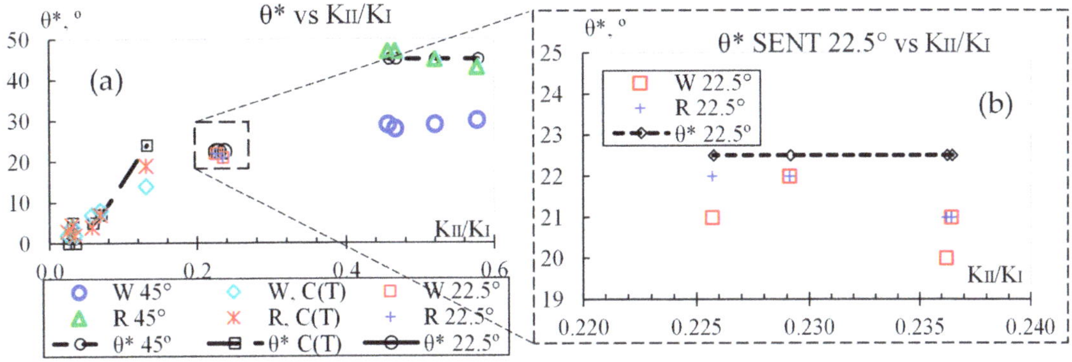

Figure 9. CP angle variation with mode mixity: (**a**) modified C(T) and SENT samples and (**b**) close-up of the SENT 22.5° sample.

On the other hand, for the modified C(T) sample, the mode mixity changes constantly, as explained in Section 4.1, and is attributed to the constant change in the stress field, starting at 0.04 and reaching 0.13. As a result, the CP for the modified C(T) sample changes as well, as seen in Figure 9 in the black squares. Thus, the mode mixity indeed influences the CP angle. This analysis shows that a change in the K_{II}/K_I ratio involves changing the CP angle.

Furthermore, in Figure 8, one can see that the size of the largest plastic zone is about 1.5 mm, so it does not exceed the holed C(T) sample thickness, 8.7 mm. On the other hand, the largest plastic zone size for the SENT sample is seen in Figure 7 for the 22.5° sample and Figure 8 for the 45° sample, where the plastic zone does not exceed the 3 mm sample thickness. Therefore, the applicability of LEFM is checked at once.

In [25], one can see how the traditional butterfly-like plastic zone in the modified C(T) evolved to be asymmetric, showing larger areas and higher stress values towards one side, the side of the stress concentrator. That stress asymmetry is assumed to cause the crack to kink. One can see in Figure 5 how the PZ turns. Although [26] did not show the evolution of the fringes for the SENT samples, they show how the fringes also exhibit asymmetry, piling up on one side. This fringe concentration reflects the stress gradient that most likely produces the crack to deviate from the direction of the applied load. The orientation change of the PZ on the C(T) or the quasi-constant orientation for the SENT is interpreted as the angle for crack kinking. Therefore, the PZ stays the same, even for the relatively large error shown by the W model. In addition, it is important to note that the C(T) sample was produced via fatigue loading, whereas the SENT sample was initially cut with a sharp razor blade and later propagated under monotonic loading. This is important because a flaw with a blunt radius does not fully comply with LEFM.

On the other hand, the CP prediction error is shown in Table 3. Both criteria are close in prediction, with errors as low as 0.28% and as high as 1.67% for the modified C(T) sample. For the largest mode mixity on the modified C(T) sample, the W criterion gives 5.56, whereas the R criterion shows a 2.78% error. The same case for the SENT 45° sample is 8.33 versus 1.11% error. In both cases, the mode mixity ($\Delta K_{II}/\Delta K_I$) had a wide range of about 0.1. For the SENT 22.5 sample, both are close to each other; the largest difference is 0.56%. In this case, this could be explained by the quasi-constant $\Delta K_{II}/\Delta K_I$.

Table 3. Error percentage in crack kinking angle prediction.

C(T)			SENT 45°			SENT 22.5°		
$\Delta K_{II}/\Delta K_I$	W, C(T)	R, C(T)	$\Delta K_{II}/\Delta K_I$	W 45°	R 45°	$\Delta K_{II}/\Delta K_I$	R 22.5°	R 22.5°
0.04	−1.11	−1.11	0.58	8.33	1.11	0.24	0.83	0.83
0.03	−1.11	−1.67	0.52	8.89	0.00	0.23	0.28	0.28
0.03	0.56	0.28	0.47	9.44	−1.11	0.24	1.39	0.83
0.06	−1.11	0.56	0.46	8.89	−1.11	0.23	0.83	0.28
0.07	−0.56	0.00						
0.13	5.56	2.78						

As a final note, it has to be said that the computational cost of these two criteria is high compared to other ones, as recently evaluated [31]. For this reason, the computational cost, in terms of floating-point operations (FLOPS), using a benchmark from [32], is compared in Table 4. The R criterion is slightly less costly than the W criterion.

Table 4. Computational cost for the evaluated criteria; baseline values from [32].

Operator	+, −, ×, /	√	Sin, Cos, Atan	Acos	Tan	ABS, SGN	^	Total Cost
Operator Cost	2	2	5	4	6	2	8	
Cost in W model	23	0	10	0	0	0	10	176
Cost in R model	33	0	5	0	0	0	8	155

5. Conclusions

The crack path was calculated for specimens under fatigue and monotonic load using the LEFM-based plastic zone radius criteria. The LEFM parameters for the fatigued sample were characterized experimentally with DIC, whereas the monotonic samples' SIFs were obtained numerically through FEM and validated with photoelasticity; for both samples, SIFs were obtained from the literature. An advantage of the plastic zone-based criteria is that the applicability of LEFM is checked in the same equation. Furthermore, using prediction models rather than FEM modeling or re-meshing makes a faster prediction by avoiding time-consuming modeling and simulation. Finally, the selected models worked for numerical and experimentally obtained SIFs. The literature has discussed how the experimentally acquired SIFs include non-linear effects such as rugosity, plasticity-induced closure, trapped debris, and crack flank interlocking.

The W and R criteria were devised for monotonic load, and here, they were tested for monotonic and fatigue load. The swap of K with ΔK was checked, which was applicable in these cases. The error analysis showed that the W and R criteria predicted angles very close for short cracks as the mode mixity ratio, $\Delta K_{II}/\Delta K_I$, is below 0.25, with the R criterion exhibiting lower error values. However, at the last measurement for the modified C(T) low carbon steel sample before unstable crack growth, the $\Delta K_{II}/\Delta K_I$ doubles its value and makes an unstable W criterion prediction. The R criterion has about 88% of the W criterion's computational cost, which could impact performance for recurrent calculations. Furthermore, the models were devised for central cracks, and here, they were tested in lateral cracks. The predicted versus experimental CP comparison showed that the models could be used for fatigue and monotonic loads.

Author Contributions: Conceptualization, J.G.D.-R.; Software, J.G.D.-R.; Writing—original draft, J.G.D.-R., O.R.B.-B. and D.P.-P.; Writing—review & editing, J.G.D.-R., A.D.P.-C. and A.M.B.B.; Formal analysis, J.G.D.-R. and D.P.-P.; Investigation, A.D.P.-C.; Methodology, A.D.P.-C.; Supervision, A.D.P.-C.; Visualization, O.R.B.-B.; Validation, O.R.B.-B.; Resources, A.M.B.B. and D.P.-P.; Funding acquisition, A.M.B.B.; Data curation, D.P.-P. All authors have read and agreed to the published version of the manuscript.

Funding: The authors acknowledge the financial support from Pontifícia Universidade Católica do Rio de Janeiro—PUC-RJ.

Data Availability Statement: Data is available upon reasonable request.

Conflicts of Interest: The authors declare no conflict of interest.

Acronyms and Nomenclature

Plastic zone	PZ
Crack path	CP
Single edge notch tension	SENT
Compact test	C(T)
Linear elastic fracture mechanics	LEFM
Floating-point operations	FLOPS
Increment in crack kinking angle	$\Delta\theta$
Stress intensity factor	SIF, K
Equivalent stress intensity factor	Keq
Stress intensity factor range factor	ΔK
Plastic zone radius	Pzr.
Fatigue threshold	ΔKth
Digital image correlation	DIC
Finite element modeling	FEM
Applied force	F
Normal stress	σ
Shear stress	τ
Radial direction	r
Tangential direction	θ
Yield stress	σ_y
Poisson's modulus	ν
Residual ligament size	(a/W)
Measure crack kinking angle	θexp
Predicted crack kinking angle	θ*

References

1. Mróz, K.P.; Mróz, Z. On Crack Path Evolution Rules. *Eng. Fract. Mech.* **2010**, *77*, 1781–1807. [CrossRef]
2. Yu, X.; Li, L.; Proust, G. Fatigue Crack Growth of Aluminium Alloy 7075-T651 under Proportional and Non-Proportional Mixed Mode I and II Loads. *Eng. Fract. Mech.* **2017**, *174*, 155–167. [CrossRef]
3. Kujawski, D. Discussion and Comments on KOP and ΔKeff. *Materials* **2020**, *13*, 4959. [CrossRef]
4. Kibey, S.; Sehitoglu, H.; Pecknold, D.A. Modeling of Fatigue Crack Closure in Inclined and Deflected Cracks. *Int. J. Fract.* **2004**, *129*, 279–308. [CrossRef]
5. Vormwald, M.; Hos, Y.; Freire, J.L.F.; Gonzáles, G.L.G.; Díaz, J.G. Crack Tip Displacement Fields Measured by Digital Image Correlation for Evaluating Variable Mode-Mixity during Fatigue Crack Growth. *Int. J. Fatigue* **2018**, *115*, 53–66. [CrossRef]
6. Maneschy, J.E.; Miranda, C.A.d.J. *Application of Fracture Mechanics in the Nuclear Industry*, 1st ed.; Eletrobras: Rio de Janeiro, Brazil, 2023; ISBN 9788599092026.
7. Worthington, M.; Chew, H.B. Crack Path Predictions in Heterogeneous Media by Machine Learning. *J. Mech. Phys. Solids* **2023**, *172*, 105188. [CrossRef]
8. Wasiluk, B.; Golos, K. Prediction of Crack Growth Direction under Plane Stress for Mixed-Mode I and II Loading. *Fatigue Fract. Eng. Mater. Struct.* **2000**, *23*, 381–386. [CrossRef]
9. Vormwald, M.; Hertel, O.; Zerres, P. Fatigue of Engineering Structures under Combined Nonproportional Loads: An Overview. *Fatigue Fract. Eng. Mater. Struct.* **2018**, *41*, 1449–1468. [CrossRef]
10. Highsmith, S. *Crack Path Determination for Non-Proportional Mixed-Mode Fatigue*; Georgia University of Technology: Atlanta, GA, USA, 2009.

11. Toribio, J.; Matos, J.C.; Gonzalez, B. Numerical Modeling of Plasticity-Induced Fatigue Crack Growth Retardation Due to Deflection in the Near-Tip Area. *Metals* **2021**, *11*, 541. [CrossRef]
12. Rozumek, D.; Macha, E. A Survey of Failure Criteria and Parameters in Mixed-Mode Fatigue Crack Growth. *Mater. Sci.* **2009**, *45*, 190–210. [CrossRef]
13. Zhu, X.K.; Joyce, J. Review of Fracture Toughness (G, K, J, CTOD, CTOA) Testing and Standardization. *Eng. Fract. Mech.* **2012**, *85*, 1–46. [CrossRef]
14. Wang, Y.; Wang, W.; Zhang, B.; Li, C.-Q. A Review on Mixed Mode Fracture of Metals. *Eng. Fract. Mech.* **2020**, *235*, 107126. [CrossRef]
15. Ren, L.; Zhu, Z.; Wang, M.; Zheng, T.; Ai, T. Mixed-Mode Elastic-Plastic Fractures: Improved R-Criterion. *J. Eng. Mech.* **2014**, *140*, 04014033. [CrossRef]
16. Richard, H.A.; Fulland, M.; Sander, M. Theoretical Crack Path Prediction. *Fatigue Fract. Eng. Mater. Struct.* **2005**, *28*, 3–12. [CrossRef]
17. Vormwald, M.; Hos, Y.; Freire, J.L.F.; Gonzáles, G.L.G.; Díaz, J.G. Variable Mode-Mixity during Fatigue Cycles—Crack Tip Parameters Determined from Displacement Fields Measured by Digital Image Correlation. *Frat. Ed Integrità Strutt.* **2017**, *11*, 314–322. [CrossRef]
18. Blanco, E.; Martínez, M.; González, J.; González, M. Fatigue Crack Growth Numerical Analysis of CPVC: Effect of Temperature and Load Frequency. *Rev. UIS Ing.* **2019**, *18*, 177–186. [CrossRef]
19. Fageehi, Y.A.; Alshoaibi, A.M. Investigating the Influence of Holes as Crack Arrestors in Simulating Crack Growth Behavior Using Finite Element Method. *Appl. Sci.* **2024**, *14*, 897. [CrossRef]
20. Berrios-Barcena, D.R.; Franco-Rodríguez, R.; Rumiche-Zapata, F.A. Calibration of Nasgro Equation for Mixed-Mode Loading Using Experimental and Numerical Data. *Rev. Fac. De Ing. Univ. De Antioq.* **2019**, *97*, 65–77. [CrossRef]
21. Shahani, A.R.; Shakeri, I.; Rans, C.D. Fatigue Crack Growth of Al 5083-H111 Subjected to Mixed-Mode Loading. *J. Braz. Soc. Mech. Sci. Eng.* **2020**, *42*, 442. [CrossRef]
22. Chatzigeorgiou, A.; Theotokoglou, E.; Tsamasphyros, G.I. Code Development for the Computational Analysis of Crack Propagation in Structures. *Frat. Ed Integrità Strutt.* **2020**, *14*, 306–324. [CrossRef]
23. Pierson, K.; Rahman, A.; Spear, A.D. Predicting Microstructure-Sensitive Fatigue-Crack Path in 3D Using a Machine Learning Framework. *JOM* **2019**, *71*, 2680–2694. [CrossRef]
24. Ricoeur, A.; Lindner, F.; Zarjov, K. Stochastic Aspects of Crack Deflection and Crack Path Prediction in Short Fiber Reinforced Polymer Matrix Composites. *Eur. J. Mech.—A/Solids* **2022**, *95*, 104598. [CrossRef]
25. Díaz-Rodríguez, J.G.; Gonzales, G.L.; Ortiz Gonzalez, J.A.; Freire, J.L.d.F. Analysis of Mixed-Mode Stress Intensity Factors Using Digital Image Correlation Displacement Fields. In Proceedings of the 24th ABCM International Congress of Mechanical Engineering, Curitiba, Brazil, 3–8 December 2017; ABCM: Curitiba, Brazil, 2017.
26. Merah, N.; Albinmousa, J. Experimental and Numerical Determination of Mixed Mode Crack Extension Angle. *J. Test. Eval.* **2009**, *37*, 95–107. [CrossRef]
27. Pook, L.P. The Effect of Crack Angle on Fracture Toughness. *Eng. Fract. Mech.* **1971**, *3*, 205–218. [CrossRef]
28. Aliha, M.R.M.; Berto, F.; Mousavi, A.; Razavi, S.M.J. On the Applicability of ASED Criterion for Predicting Mixed Mode I+II Fracture Toughness Results of a Rock Material. *Theor. Appl. Fract. Mech.* **2017**, *92*, 198–204. [CrossRef]
29. Miranda, A.C.O.; Meggiolaro, M.A.; De Castro, J.T.P.; Martha, L.F. Fatigue Life Prediction of Complex 2D Components under Mixed-Mode Variable Amplitude Loading. *Int. J. Fatigue* **2003**, *25*, 1157–1167. [CrossRef]
30. Sutton, M.A.; Orteu, J.J.; Schreier, H. *Image Correlation for Shape, Motion and Deformation Measurements: Basic Concepts, Theory and Applications*; Springer: Boston, MA, USA, 2009; ISBN 978-0-387-78746-6.
31. Díaz, J.G.; Freire, J.L.d.F. LEFM Crack Path Models Evaluation under Proportional and Non-Proportional Load in Low Carbon Steels Using Digital Image Correlation Data. *Int. J. Fatigue* **2022**, *156*, 106687. [CrossRef]
32. Atkinson, L. A Simple Benchmark of Various Math Operations. Available online: https://latkin.org/blog/2014/11/09/a-simple-benchmark-of-various-math-operations/ (accessed on 2 November 2023).

buildings

MDPI

Article

Parallelization Strategy for 3D Probabilistic Numerical Cracking Model Applied to Large Concrete Structures

Mariane Rodrigues Rita [1], Pierre Rossi [1,*], Eduardo de Moraes Rego Fairbairn [1], Fernando Luiz Bastos Ribeiro [1], Jean-Louis Tailhan [2], Henrique Conde Carvalho de Andrade [1] and Magno Teixeira Mota [1]

1 Civil Engineering Program, Universidade Federal do Rio de Janeiro (UFRJ), Av. Horácio Macedo, 2030, Cidade Universitária, Rio de Janeiro 21941-598, Brazil; mariane_rita@coc.ufrj.br (M.R.R.); eduardo@coc.ufrj.br (E.d.M.R.F.); fernando@coc.ufrj.br (F.L.B.R.); henriqueconde@coc.ufrj.br (H.C.C.d.A.); magnomota@coc.ufrj.br (M.T.M.)

2 Department of Materials and Structures, Gustave Eiffel University, 13015 Marseille, France; jean-louis.tailhan@univ-eiffel.fr

* Correspondence: p.lupo@yahoo.fr

Abstract: This work presents the application of a finite element model utilizing a three-dimensional (3D) probabilistic semi-explicit cracking model to analyze the rupture process of a large concrete wall beam. The numerical analysis predicts both the global behavior of the structure and its primary rupture mechanisms, utilizing three different finite element mesh refinements to ensure robustness. A Monte Carlo (MC) procedure is integrated into the modeling approach to account for probabilistic variations of the material properties. The statistical analysis derived from this probabilistic model may sometimes result in overly conservative safety coefficients, particularly when using a coarse mesh. Additionally, the detailed understanding of the structure's cracking process, regardless of its rupture mechanism, may experience some reduction in precision. Due to the necessity of numerous simulations to achieve statistically significant results, the MC procedure can become computationally expensive. To address this, a straightforward parallelization of the Monte Carlo procedure was implemented, allowing multiple finite element analyses to be conducted concurrently. This strategy significantly reduced computational time, thereby enhancing the efficiency of the numerical model in performing numerical simulations of structural engineering.

Keywords: concrete structures; probabilistic numerical model; macrocracks pattern; monte carlo method; parallelization strategy

1. Introduction

The use of computational parallelization strategies has become increasingly crucial in solving complex engineering problems, especially when a large number of equations need to be solved [1–6]. This is particularly true for the probabilistic models applied to the 3D simulation of concrete structures, which require numerous computational analyses due to the use of random distributions of material properties to represent heterogeneity [7–9]. In this context, computational parallelization facilitates more robust and detailed simulations by allowing the simultaneous execution of multiple analyses, significantly reducing processing time and increasing simulation efficiency.

Furthermore, the development of accurate numerical models for analyzing the rupture process in large concrete structures plays a pivotal role in ensuring their structural integrity and safety. A critical phenomenon that influences a structure's response based on its size or volume, is the scale effect. This effect is attributed to concrete's heterogeneous nature, physical and chemical phenomena during production, drying-induced micro-cracking, and water-induced porosity. Therefore, considering that the scale effect in the modeling of concrete structures is primordial due to the heterogeneity of the material, addressing it involves understanding both material-level constitutive relations and structure-level design methods and finite element analysis.

In this context, this work aims to analyze the cracking process in a very large concrete structure utilizing a three-dimensional (3D) probabilistic semi-explicit cracking model that has been developed and detailed in previous studies [8,10–12]. The model validation was conducted through the simulation of a fracture mechanical test using a large Double Cantilever Beam (DCB) specimen. The dimensions of this DCB specimen were significant, measuring 3.5 m in length, 1.1 m in width, and 0.3 m in height. Notably, the observed and modeled crack propagation length exceeded 2 m [13].

The structural case analyzed here is a concrete wall beam that can present more than one macrocrack unlike to the single macrocrack scenario observed in the DCB specimen [12], providing a more realistic representation of a structural behavior. To ensure the robustness of the model, three mesh refinements are employed to verify the numerical results independence of the finite element mesh.

The probabilistic numerical model focuses specifically on the propagation of macrocracks within concrete structures. To simulate the concrete heterogeneous behavior within the finite element method framework, it is necessary to use a Monte Carlo (MC) approach. Due to the need to use several samples to obtain statistically consistent results, the execution of the MC procedure becomes computationally very time-consuming, depending on the problem at hand. Therefore, given the computational cost of these simulations, optimizing the calculation time is critical.

As a solution to overcome this issue, a straightforward parallelization of the Monte Carlo procedure was implemented, allowing for multiple finite element analyses to be performed simultaneously. The study case demonstrates that the use of this strategy significantly reduces computational time, allowing the practical applicability of probabilistic numerical models in the analysis and design of real-world engineering problems of concrete structures.

The decision to enhance execution time by implementing the Monte Carlo (MC) method in parallel, rather than pursuing parallelization of both Finite Element Method (FEM) and MC codes, stems from some key factors. Firstly, the parallel implementation of the MC method offers simplicity in execution. Additionally, since FEM analyses are inherently independent tasks, there is no requirement for extensive communication overhead. This approach facilitates the concurrent execution of multiple FEM analyses on the computational platform, resulting in significant runtime reductions. Nonetheless, it is worth noting that future endeavors may explore enhancements to this strategy, including the investigation of dual-code parallelization involving both FEM and MC codes.

2. Principles of the Semi-Explicit Probabilistic Cracking Model

This study introduces a 3D semi-explicit probabilistic model for crack propagation, previously detailed in [12,14]. The model focuses specifically on analyzing crack propagation in concrete structures under mode I failures induced by tensile stresses. Developed using the FEM, the model incorporates considerations of heterogeneity and volume effects within a probabilistic framework, crucial for accurately simulating crack propagation in concrete structures. The model is implemented in FORTRAN [15,16] and integrated into a versatile FEM platform created by researchers from the Civil Engineering Program at COPPE/UFRJ. This platform has been widely applied across various studies, demonstrating its utility in diverse applications [1,8,9,17].

To address material characteristics such as heterogeneity and scale effects, the numerical model assumes that each finite element represents a distinct volume of heterogeneous material. This material's behavior is influenced by its degree of heterogeneity (r_e), and determined by the ratio of the finite element volume (V_e) to the volume of the coarsest aggregate (V_a):$(r_e = V_e/V_a)$. The model adopts a volume element-based approach to simulate macrocrack propagation within concrete structures. The criteria governing macrocrack propagation can be divided into two aspects: macrocrack initiation and macrocrack propagation. Macrocrack initiation depends directly on the uniaxial tensile strength (f_t) of the concrete material. Once initiated, macrocrack propagation is guided by the dissipation of post-cracking energy in tension, ensuring the model accurately represents the mechanics of crack propagation in concrete structures. A macrocrack is considered fully propagated once all post-cracking dissipation energy is consumed.

An innovative feature of the model is its utilization of Linear Elastic Fracture Mechanics (LEFM) to derive post-cracking dissipation energy. This approach employs the mode I critical fracture energy G_{IC} to quantify the energy dissipated during crack propagation. Both the uniaxial tensile strength and mode I critical fracture energy are treated as probabilistic mechanical characteristics within the model. The random distribution of f_t follows the Weibull distribution [18,19], while G_{IC} follows a lognormal distribution [20]. Accurately defining the parameters of these distributions is crucial for accurately representing material behavior.

At the microscale within each finite element (FE) volume, the cracking process results in energy dissipation governed by an isotropic damage law, as depicted in Equation (1), where the variables $\tilde{\varepsilon}_0$, $\tilde{\varepsilon}_{fi}$, and $\tilde{\varepsilon}^k$, respectively, represent the damage initiation strain, maximum critical strain, and equivalent strain. This dissipative process initiates when the maximum principal stress (σ_1) reaches the randomly assigned concrete tensile strength (f_t). Thereafter, once the total available energy for the FE is consumed, the element is considered cracked, and its stiffness matrix is considered null [11].

$$D = 1 - \frac{\tilde{\varepsilon}_0}{\tilde{\varepsilon}^k}\left[1 - \frac{(\tilde{\varepsilon}^k - \tilde{\varepsilon}_0)}{(\tilde{\varepsilon}_{fi} - \tilde{\varepsilon}_0)}\right] \quad (1)$$

An illustration of the main principles of the proposed model is depicted in Figure 1. This figure highlights the random distribution of concrete's tensile strength and the volumetric density of dissipated energy, represented as g_{IC}, along with the elementary representation of the damage evolution process. Each finite element follows unique laws governing this process due to inherent randomness. The determination of g_{IC} utilizes an energetic regularization technique [21], which relates the material fracture energy G_{IC} to the elementary characteristic length l_e through the equation: $g_{IC} = G_{IC}/l_e$, where $l_e = (V_e)^{1/3}$.

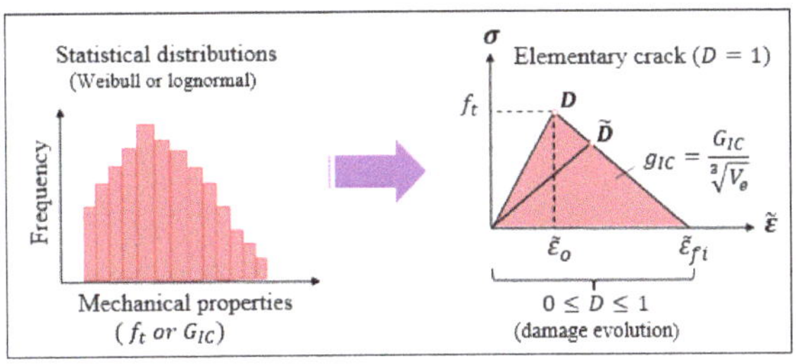

Figure 1. Illustration of the damage evolution at the finite element level and random mechanical properties distribution.

In this modeling approach, it is important to emphasize cracks emerge and propagate at a macroscopic scale due to the elemental failure of successive elements, which occur randomly and may coalesce to form macroscopic cracks. A finite element (FE) is considered cracked only when its associated energy is fully dissipated. Therefore, within this framework, the model does not address traditional crack propagation laws as seen in classical damage model or smeared crack model [22–28], as the inclusion of the damage parameter in the constitutive model solely serves to dissipate G_{IC}.

Moreover, in this approach, each finite element analysis acts as a sample within the Monte Carlo procedure, yielding global structural responses of the model. This methodology involves executing multiple FEM simulations for a given structural scenario, incorporating diverse spatial distributions of mechanical material properties to ensure statistically significant outcomes.

2.1. Model's Parameters Estimation

Accurately determining parameters for the Weibull and lognormal distributions is crucial to ensure the model's consistency. The Weibull distribution is characterized by shape and scale parameters (b, c), while the lognormal distribution is defined by mean and standard deviation parameters (μ, σ). These parameters govern the stochastic distribution of tensile strength and fracture energy, capturing the heterogeneous behavior of materials at the finite element scale.

Fracture energy (G_{IC}) is assumed to be an intrinsic property of the material with a constant mean value $\mu(G_{IC})$, independent of scale and specific to the type of concrete. The standard deviation $\sigma(G_{IC})$ of the lognormal distribution thus becomes a key parameter to be determined. This assumption aligns with findings from Rossi [13], acknowledging inherent material variability. Therefore, the primary parameters to be determined are as follows: (b, c) for the Weibull distribution and $\sigma(G_{IC})$ for the lognormal distribution. Recent advancements [12,14] now enable the estimation of these model parameters using concrete compressive strength, maximum aggregate volume, and finite element mesh size data. This capability significantly enhances the model's versatility and applicability across diverse concrete compositions.

The dependency of $\sigma(G_{IC})$ on finite element volume was a significant outcome in the model's development [8,12]. This relationship was established through inverse analysis of a concrete mix design, validated by simulating a macrocrack propagation test on a large DCB specimen and comparing results with experimental data from Rossi [13]. An equation was derived correlating the lognormal distribution's standard deviation with mesh heterogeneity. Subsequently, the application of the model was extended to include other concrete mixes with compressive strengths below 130 MPa and aggregate diameters of 10 mm to 20 mm, by proposing a methodology to estimate the parameters of G_{IC} [14].

For the Weibull distribution parameters, an iterative numerical method was developed. This method integrates expressions for the mean tensile strength $\mu(f_t(r_e))$ and coefficient of variation $\frac{\sigma}{\mu}(f_t(r_e))$, derived from experimental data and scaling laws. It estimates (b, c) parameters for specific concrete volumes, validated for compressive strengths up to 130 MPa and maximum aggregate sizes of 10 mm or more. Applied directly at the finite element level, this method assigns (b, c) parameters to each element based on its volume, maximum aggregate size, and compressive strength (f_c).

2.2. Parallelization of the Monte Carlo Method

In this study, a Monte Carlo procedure is used for simulating probabilistic concrete cracking, introducing stochasticity at the material's local scale. Each MC sample corresponds to a finite element problem solution, treated as an independent sample. Random variables are generated for each sample based on the probability density functions, yielding load-displacement diagrams (P-δ) as the outcome.

The computational demands of the MC procedure increase notably due to each sample involving solving a finite element problem. To address this, a parallel programming strategy was employed to implement the MC method using the OpenMP application programming interface. OpenMP facilitates the development of multi-threaded applications in languages such as C/C++ and FORTRAN. Comprising compilation directives, function libraries, and environment variables, OpenMP supports both explicit and implicit multi-threaded parallelism, making it a versatile tool for parallel computing [29–31].

The parallel algorithm for the Monte Carlo method employs a master-slave parallel model, which is widely utilized in networked computing environments [32]. In this approach, the code initiates with a master thread that executes sequentially until entering a parallel region. During this phase, the master thread spawns multiple parallel threads, each executing tasks concurrently within the parallel region. Upon the completion of tasks, the threads synchronize and terminate, leaving only the master thread. This parallel model aligns with the sequential algorithm's principles, facilitating the simultaneous execution of multiple independent FEM analyses. The main steps of the Parallel Monte Carlo (PMC) procedure are outlined in Algorithm 1. This method was selected for its straightforward implementation and compatibility with the stochastic nature of the proposed methodology, which necessitates a large number of independent simulations.

Algorithm 1 Parallel Monte Carlo Procedure

```
Variables initialization
Read maxmc and stop criteria ; // maxmc is the maximum number of samples
Read nthreads ; // nthreads is the number of threads
j = 0 ; // counter of performed samples
while stopping criterion is not reached do
    j = j + nthreads
    Initialize the parallel environment
        For each block with nthreads samples do
            Perform the FEM analysis
            Save (P, δ) curve and relevant results of the sample
        end for
    Finalize the parallel environment
end while
Perform the statistical analysis
```

The selected stopping criterion for PMC simulations was reaching the predefined number of samples. Because FEM analyses differ due to the problem's probabilistic nature, tasks are unevenly balanced. Thus, a dynamic schedule clause was utilized in PMC implementation, distributing tasks based on execution time, with new tasks assigned as threads become idle.

3. Structural Case and Numerical Simulations

The structural case investigated in this study involves a sizable wall beam subjected to four-point bending, characterized by dimensions of 5 m in length, 3 m in height, and 0.5 m in thickness. A representation of the test configuration is shown in Figure 2, where P is the total force applied to the specimen by two loading pins; d is the distance between the supporting and loading pins; l is the specimen width, and h is the specimen height. The distance between the bottom loading supports is 5 m and between the upper loading supports is 2.2 m.

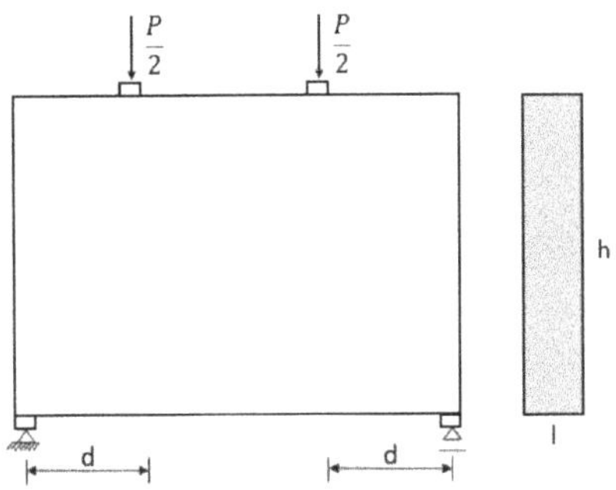

Figure 2. Representation of the four-point bending test.

In contrast to common practices involving large reinforced concrete beams, often with or without fibers, the beam used as the structural case in this study comprises plain concrete without reinforcements. The primary objective of the numerical simulations is to demonstrate the model's applicability. Through comparisons across different levels of mesh refinement, the study aims to highlight the model's robustness and independence from mesh density. Additionally, the investigation seeks to validate the efficacy of the estimated probabilistic parameters.

In the numerical simulations, three-dimensional finite element analyses were conducted using three distinct mesh refinements. The refinement level of each mesh is determined by the parameter r_e^{mean}, calculated as V_e^{mean}/V_a, where V_e^{mean} represents the mean volume of the finite elements within each mesh. The meshes consist of tetrahedral elements approximated by linear interpolation functions. Visual representations of the three finite element meshes can be found in Figures 3–5, while detailed information regarding the meshes, including the number of elements and nodes, as well as the maximum and minimum volumes of the finite elements, is provided in Table 1.

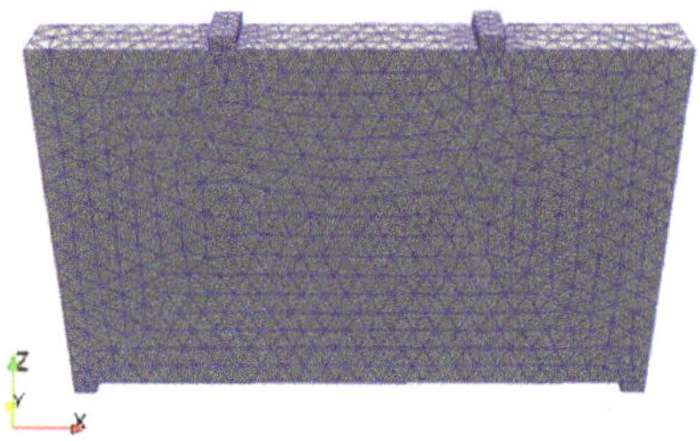

Figure 3. Three-dimensional Mesh 1 with $r_e^{mean} \approx 829$.

Figure 4. Three-dimensional Mesh 2 with $r_e^{mean} \approx 223$.

Figure 5. Three-dimensional Mesh 3 with $r_e^{mean} \approx 74$.

Table 1. Description of the three finite element meshes.

Reference	Neq	Nnode	Numel	r_e^{mean}	r_e^{sd}	r_e^{min}	r_e^{max}
Mesh 1	7180	2403	10,364	829.1	276.3	110.6	2963.9
Mesh 2	22,756	7595	37,503	223.3	63.1	54.4	710.5
Mesh 3	64,243	21,424	112,572	73.9	22.3	14.4	207.4

The mechanical characteristics of concrete, which are identical to those of the DCB previous modeling [8,14], are the following: compressive strength f_c = 54 MPa; splitting tensile strength f_t = 4.1 MPa; Young modulus E = 35.5 MPa; fracture energy G_{IC} = 1.25 × 10^{-4} MN/m. The maximum aggregate size is 12 mm, corresponding to $V_a = 0.9$ cm^3. Across all simulations, a total displacement of 2.5 mm is applied, divided into 50 steps of 0.05 mm each. Additionally, thirty samples are analyzed for each Monte Carlo simulation.

3.1. Numerical Results

All results are presented in terms of the global behavior of the structure, specifically load-displacement curves depicting the wall beam's overall mechanical response. The Monte Carlo simulation results for the three mesh refinements are displayed in Figures 6–8, encompassing 30 finite element analyses in each case. Figure 9 shows a comparison between the arithmetic mean curves obtained for each mesh refinement. As can be observed, the load-displacement curves for the three meshes exhibit similar behavior, particularly in the cases of Mesh 2 and Mesh 3, indicating coherence in determining the statistical parameters.

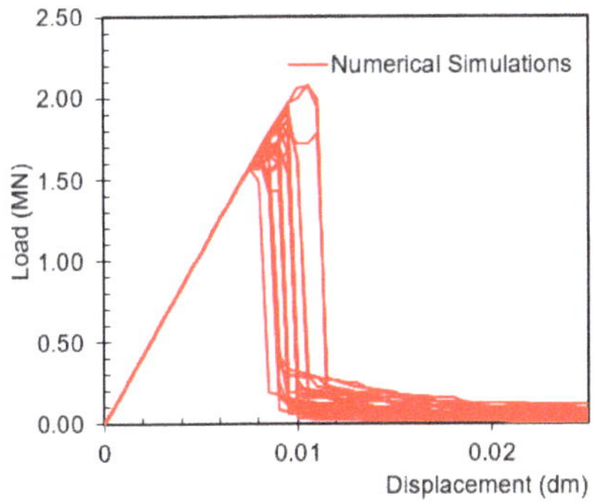

Figure 6. Load-displacement curves obtained from Monte Carlo samples for Mesh 1 ($r_e^{mean} \approx 829$).

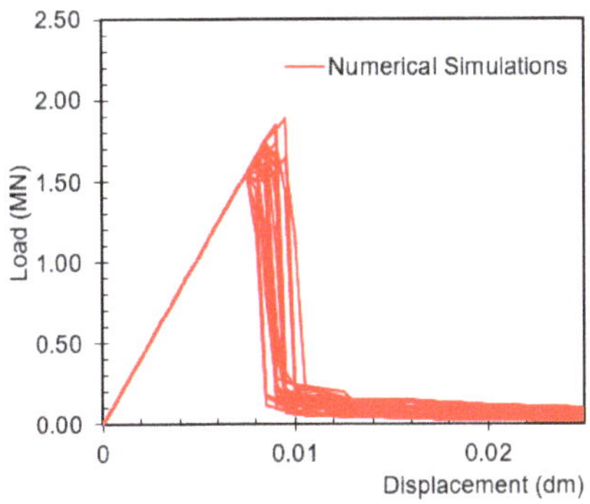

Figure 7. Load-displacement curves obtained from Monte Carlo samples for Mesh 2 ($r_e^{mean} \approx 223$).

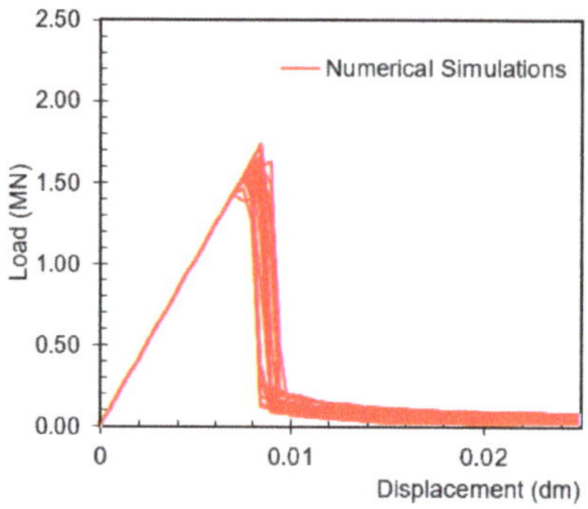

Figure 8. Load-displacement curves obtained from Monte Carlo samples for Mesh 3 ($r_e^{mean} \approx 74$).

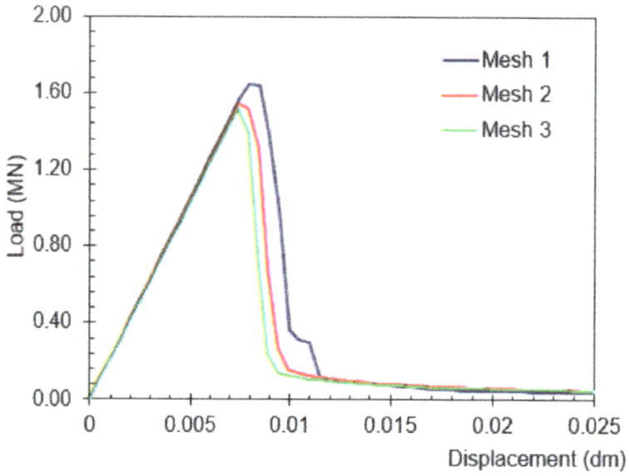

Figure 9. Comparison between the mean structural global response of the three mesh refinements.

3.2. Analysis of the Results

Upon examining Figures 6–8, it becomes evident that the dispersion associated with the global behavior of the structure decreases as the mesh is refined. Statistically, this suggests that using Mesh 1 (the coarsest option) would yield a larger safety coefficient compared to Mesh 2 and especially Mesh 3.

From the analysis of the results presented in Figure 9 the following can be stated: (1) All three meshes exhibit a similar global behavior for the structure. This brittleness aligns well with the nature of the considered structural problem, given that the structure lacks reinforcement; (2) The mesh dependency remains minimal, even with a substantial difference in refinements between Meshes 1 and 3 (there is practically no difference in behavior between meshes 2 and 3).

Examining Figures 10–12 it can be observed that the primary crack mechanism leading to the structural rupture remains consistent across all three meshes, manifesting as a macrocrack originating from the left bottom loading support and extending to the left upper loading support. However, the detailed description of the cracking pattern diverges between Mesh 1 and Meshes 2 and 3. With Mesh 1, only a single macrocrack is observed. In contrast, Meshes 2 and 3 exhibit a more intricate pattern, featuring a second macrocrack initiating from the right bottom loading support.

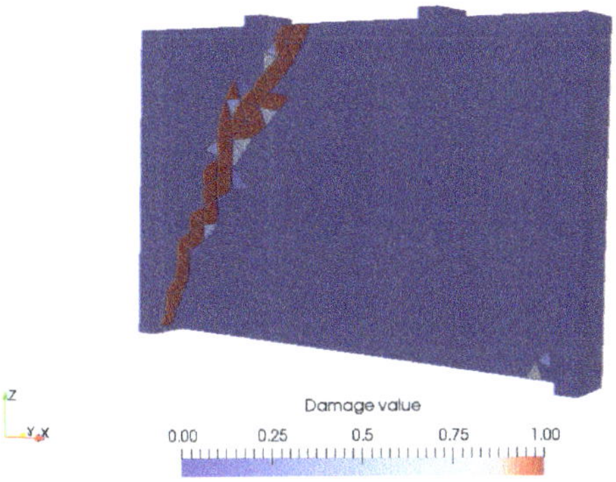

Figure 10. Example of cracking pattern for sample 21 of Mesh 1 ($r_e^{mean} \approx 829$).

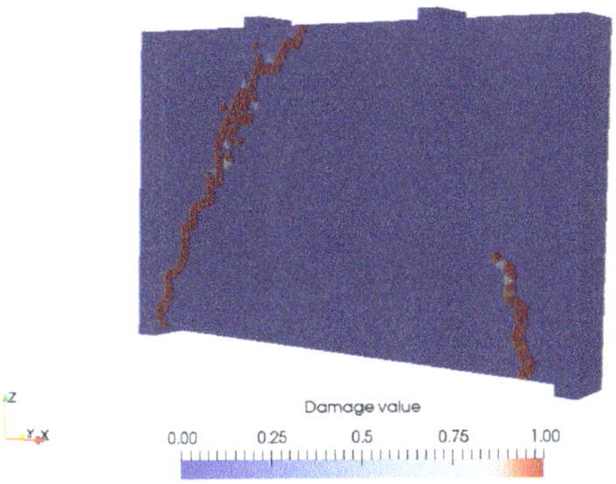

Figure 11. Example of cracking pattern for sample 17 of Mesh 2 ($r_e^{mean} \approx 223$).

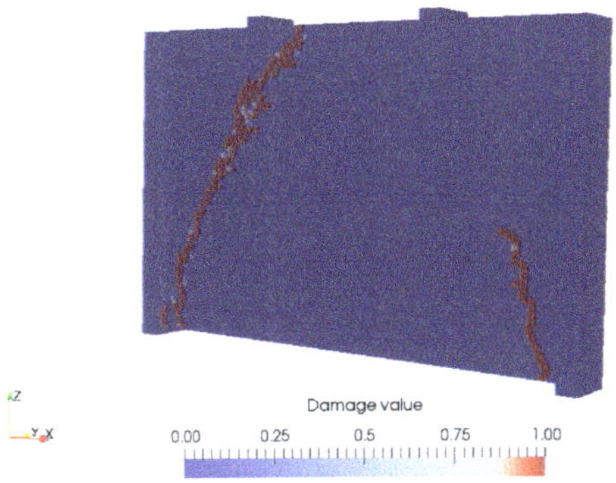

Figure 12. Example of cracking pattern for sample 14 of Mesh 3 ($r_e^{mean} \approx 74$).

In summary, it can be affirmed that the refinement of Mesh 1 is sufficient to yield an acceptable global behavior of the structure, possibly with an overly conservative safety coefficient. It also facilitates a proper rupture mechanism. Nevertheless, it does not attain the level of fineness required for a detailed analysis of the cracking process.

4. Parallelization Performance Analysis

A parallel processing environment typically consists of multiple processors with shared or distributed memory, an operating system for managing processes, and a parallel algorithm [33]. In distributed memory systems, each processor has its local memory, and data is exchanged via a high-speed network using a message system like MPI (Message Passing Interface). Shared memory systems feature multiprocessors that share a single memory space, communicating through shared variables. Hybrid systems combine both forms, with nodes that have separate address spaces and multiple processors sharing memory within each node [34].

In this study, performance simulations are conducted on a parallel computing platform consisting of a cluster comprising four nodes interconnected by a local Ethernet network with a speed of 10 Gb/s. Each node is equipped with dual processors utilizing a cache-coherent Non-Uniform Memory Access (cc-NUMA) architecture [35]. Detailed specifications of the hardware for each node can be found in Table 2. Communication between nodes on this platform occurs via the local network, while intra-node communication between processors is managed by a Quick Path Interconnect (QPI) circuit [3].

Table 2. General characteristics of each node of the cluster.

Characteristic	**Description**
Processor	Intel Xeon E5-2630 v4
Number of processors	2
Cores	2 × 10
Socket per node	2
Number of threads	20
Cache size (L3)	25 MB
Clock	2.2 Ghz
Memory per node	256 GRAM (DDR4-2134 MHz)
Motherboard	Intel S2600CWR
LAN	Gigabit (2 × 10 Gbit)
Operational System	CentOS 7

There are several metrics available to gauge the performance of parallel applications, the most common being speedup, efficiency, and FLOPs (floating-point operations per second). In this work, speedup is selected as the primary metric due to its practicality and clarity in demonstrating how much faster the parallel code runs compared to its sequential counterpart. Speedup (denoted as S_n) is defined as the ratio of the sequential execution time T_s to the parallel execution time T_n. Mathematically, S_n is calculated as:

$$S_n = \frac{T_s}{T_n} \tag{2}$$

It is noteworthy that the execution times reported in this work refer to the wall time, which is the actual elapsed time of the program's execution. This time measurement is preferable to CPU time for evaluating parallel performance, and therefore, it is adopted in this work.

4.1. Description of the Performance Analysis

The parallelization strategy was designed with the understanding that each MC sample corresponds to an independent FEM analysis, while the probabilistic model response relies on multiple analyses of the problem. Hence, a direct parallelization approach was adopted, enabling the concurrent execution of multiple finite element analyses. During performance tests, simulations of parallel Monte Carlo were conducted with each Monte Carlo simulation (set of samples) running on a separate node of the cluster, thereby constituting shared memory tests.

The number of samples in each MC simulation during performance tests is set to twenty ($n_{\mathrm{mc}} = 20$) to align with the available threads per cluster node, optimizing speedup test duration. Consequently, the first twenty FE analyses per problem solution are used for each case. Consistency is ensured by using saved random seeds for initializing random property generation, thereby maintaining identical scenarios for precise runtime comparisons.

The parallel computational platform, consisting of four nodes, enabled the simultaneous execution of up to four problems throughout the performance analysis phase. This capability provided a significant advantage for executing the inverse analysis procedure to estimate model parameters and evaluate speedup tests. However, since these cases were independently performed on each cluster node utilizing available hardware resources, a specific study quantifying the time saved by concurrently using all four nodes was not conducted, despite the intuitive notion of time savings.

4.2. Performance Results

The results of parallelizing the Monte Carlo procedure are illustrated in Figures 13–16. Figure 13 provides a comparison of the speedup obtained for each mesh refinement, while Figures 14–16 show (a) the speedup measures, and (b) the execution time. The speedup is measured relative to the number of threads (*nt*) used, with *nt* taking the values of 1, 2, 5, 10, and 20.

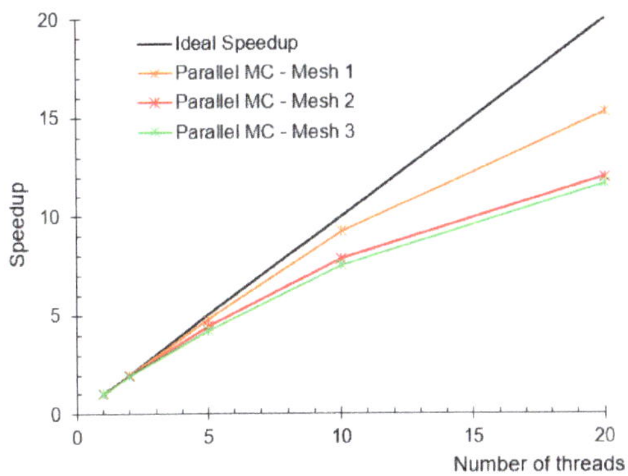

Figure 13. Comparison between the speedups of the three wall beam meshes.

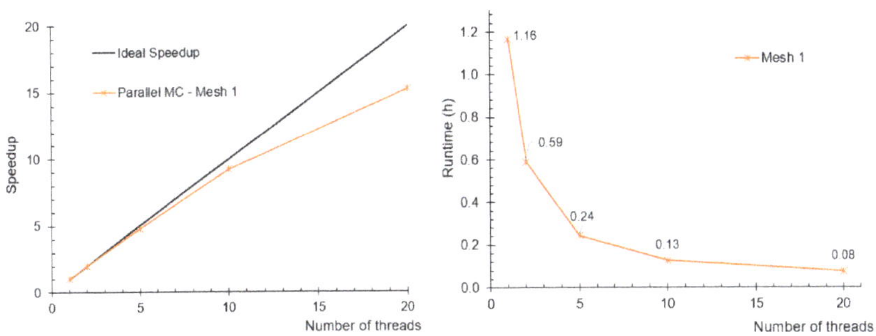

(**a**) Speedups. (**b**) Simulation runtime.

Figure 14. Performance results obtained with PMC—wall beam Mesh 1.

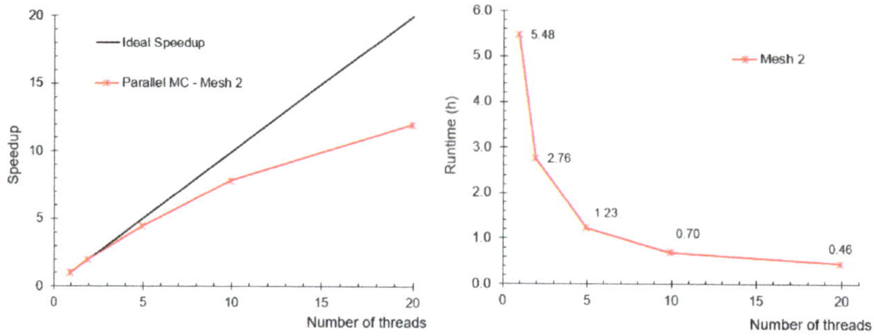

(a) Speedups. **(b)** Simulation runtime.

Figure 15. Performance results obtained with PMC—wall beam Mesh 2.

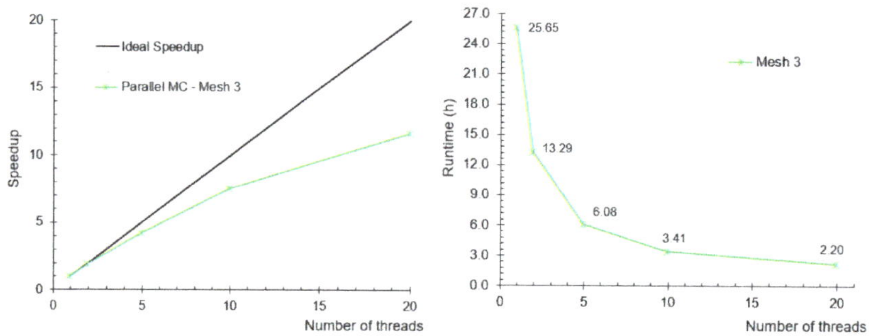

(a) Speedups. **(b)** Simulation runtime.

Figure 16. Performance results obtained with PMC—wall beam Mesh 3.

The results indicate that Mesh 1 exhibits superior speedup compared to Meshes 2 and 3. Meshes 2 and 3 demonstrate comparable outcomes, though consistently inferior to Mesh 1. The overall reduction in execution time, considering both sequential and 20-thread runtimes, is as follows for the three cases: Mesh 1 decreases from 1.16 h to 0.08 h; Mesh 2 decreases from 5.48 h to 0.46 h and Mesh 3 decreases from 25.65 h to 2.2 h.

Discussions and Remarks

Based on the reported results, it is evident that the PMC has notably decreased the execution time of the simulated cases. Concerning the speedup measures, it can be seen that as the number of threads increases, the speedup values start to deviate from the ideal speedup curve. This difference becomes more pronounced for Mesh 1 when 20 threads are used (15.27) whereas, for Meshes 2 and 3, it occurs as early as 10 threads (7.86 and 7.52, respectively). The values of speedup achieved with 20 threads for Meshes 2 and 3 are similar, 11.99 and 11.66, respectively. It is important to mention that since each sample's runtime varies depending on the distribution of elementary random parameters, a loss of efficiency may occur when there is a significant difference between the runtimes of different samples, despite using dynamic scheduling in the parallel region implementation.

To assess the impact of varying sample execution times on parallelization performance, additional Monte Carlo (MC) simulations were conducted. These simulations focused on an illustrative scenario where all MC samples are identical, sharing the same elementary properties. The aim of these hypothetical cases is to determine the maximum potential gain achievable through parallelization. The results of the speedups obtained for these new MC simulations are presented in Table 3, where *nt* represents the number of threads used.

The results are categorized as "distinct samples", reflecting those previously reported in Section 4.2, and "equal samples", representing the hypothetical scenario described here.

Table 3. Comparison between the speedups obtained for the real cases and the hypothetical cases.

Mesh Reference	nt	Distinct Samples	Equal Samples
Mesh 1	1	1.00	1.00
	2	1.97	2.05
	5	4.76	5.10
	10	9.23	10.33
	20	15.27	19.86
Mesh 2	1	1.00	1.00
	2	1.98	2.02
	5	4.44	5.04
	10	7.86	9.94
	20	11.99	16.29
Mesh 3	1	1.00	1.00
	2	1.93	1.96
	5	4.22	4.65
	10	7.52	8.93
	20	11.66	13.87

As can be observed, the use of identical samples enhances parallelization performance across all analyzed scenarios. The performance improvement percentages for the three mesh refinements, when utilizing 20 threads, are 30.04%, 35.83%, and 18.92%, respectively. For this illustrative scenario, the best speedup is obtained for Mesh 1 (19.86) and it nearly reaches the ideal speedup value when 20 threads are used. For Mesh 2, the performance gain is closer to the ideal speedup up to the use of 10 threads, with a deviation of the ideal value when 20 threads are used. Conversely, for Mesh 3, the speedup values start to deviate from the ideal speedup from the use of 10 threads.

As evident, employing identical samples significantly enhances parallelization performance across all analyzed mesh refinements. The performance improvement percentages for the three mesh refinements when utilizing 20 threads are 30.04%, 35.83%, and 18.92%, respectively. In this illustrative scenario, the most substantial speedup achieved with 20 threads is for Mesh 1 (19.86), nearly reaching the ideal speedup. Additionally, a super-linear speedup is observed when 10 threads are used (10.33). For Mesh 2, the results deviate from the ideal speedup with 20 threads (16.29), yet exhibit a slight super-linear speedup with five threads (5.04) and approximate the ideal with 10 threads (9.94). Conversely, for Mesh 3, speedup values begin deviating from the ideal with 10 threads (8.93), with a more pronounced impact when 20 threads are employed (13.87). In all cases, it is observed that a speed-down when all cores are used. This behavior may stem from cache memory access saturation due to the utilization of all cores [3].

5. Conclusions

In summary, this study applies a 3D probabilistic semi-explicit cracking model to analyze macrocrack propagation in large structures using a numerical simulation of a wall beam as a case study. This structural scenario illustrates situations where more than one macrocrack can develop. The results demonstrated mesh independence and underscored the model's reliability in predicting both global behavior and primary rupture mechanisms within structures. While the model showcases robustness, it is crucial to exercise caution with coarser meshes, as they may yield conservative safety coefficients and reduce the precision of the detailed cracking process, irrespective of rupture mechanisms.

The incorporation of parallel Monte Carlo into the probabilistic approach has significantly reduced simulation times across diverse mesh configurations. Despite its simplicity, this integration represents a substantial advancement within the framework of this probabilistic approach. Notably, for Mesh 1, runtime decreased from 1.16 h to 0.08 h; for Mesh 2, from 5.48 h to 0.46 h and for Mesh 3, from 25.65 h to 2.2 h. This advancement is crucial for facilitating extensive simulations required by this modeling approach, marking a significant stride toward applying the probabilistic model effectively in three-dimensional analyses for large-scale structures.

Looking ahead, further refinement of the model is essential for practical applications in real-world large concrete constructions, particularly in modeling steel rebars and interactions between concrete and steel. Additionally, the model should be expanded to simulate macrocrack propagation in fiber-reinforced concrete, thereby broadening its applicability in structural analysis and design to encompass more comprehensive scenarios.

Author Contributions: Conceptualization, M.R.R., P.R., J.-L.T., E.d.M.R.F. and M.T.M.; methodology, P.R., J.-L.T. and E.d.M.R.F.; software, M.R.R., F.L.B.R. and H.C.C.d.A.; validation, M.R.R.; formal analysis, M.R.R.; writing—original draft preparation, M.R.R. and P.R.; writing—review and editing, M.R.R., P.R., E.d.M.R.F. and H.C.C.d.A.; supervision, E.d.M.R.F., J.-L.T., P.R. and F.L.B.R. All authors have read and agreed to the published version of the manuscript.

Funding: This study was financed by Brazilian scientific agencies, namely the Conselho Nacional de Desenvolvimento Científico e Tecnológico (CNPq), and the Coordenação de Aperfeiçoamento de Pessoal de Nível Superior (CAPES)—Finance Code 001.

Data Availability Statement: The data presented in this study are available in the article.

Conflicts of Interest: The authors declare no conflict of interest.

References

1. Ferreira, I.A. Solução em Paralelo de um Modelo Termo-Químico-Mecânico para Concreto Jovem. Ph.D. Thesis, UFRJ, Rio de Janeiro, Brazil, 2008.
2. Krejsa, M.; Cajka, P.J.R.; Brozovsky, J.; Krejsa, V. *Insights and Innovations in Structural Engineering, Mechanics and Computation*; CRC Press/Balkema: Leiden, The Netherlands, 2016; Chapter Probabilistic Structural Modelling in Parallel Systems.
3. Andrade, H.C.C. Modelagem de Escoamentos Reativos Levemente Compressíveis com um Modelo Computacional Paralelo de Volumes Finitos. Ph.D. Thesis, Universidade Federal do Rio de Janeiro, Rio de Janeiro, Brazil, 2020.
4. Na, O.; Xi, Y. Parallel Finite Element Model for Multispecies Transport in Nonsaturated Concrete Structures. *Materials* **2019**, *12*, 2764. [CrossRef] [PubMed]
5. Zheng, X.; Jin, J.; Wang, Y.; Yuan, M.; Qiang, S. Research on the Application and Performance Optimization of GPU Parallel Computing in Concrete Temperature Control Simulation. *Buildings* **2023**, *13*, 2657. [CrossRef]
6. Martins, M.A.D.; Alves, J.L.D.; Coutinho, A.L.G.A. Parallel Edge-Based Finite Element Techniques for Nonlinear Solid Mechanics. In Proceedings of the Vector and Parallel Processing—VECPAR 2000, Porto, Portugal, 21–23 June 2000; Palma, J.M.L.M., Dongarra, J., Hernández, V., Eds.; Springer: Berlin/Heidelberg, Germany, 2001; pp. 506–518.
7. Mazars, J. Application de la Mecanique de L'endommagement au Comportement non Lineaire et a la Rupture du Beton de Structure. Ph.D. Thesis, Ecole Normale Supériore de l'Enseignement Technique, Paris, France, 1984.
8. Rita, M.R. Implementation of a 3D Macroscopic Probabilistic Model for Semi-Explicit Concrete Cracking. Ph.D. Thesis, UFRJ, Rio de Janeiro, Brazil, 2022.
9. Mota, M.; Fairbairn, E.; Ribeiro, F.; Rossi, P.; Tailhan, J.L.; Andrade, H.; Rita, M. A 3D probabilistic model for explicit cracking of concrete. *Comput. Concr.* **2021**, *27*, 549–562.
10. Rossi, P.; Wu, X.; Le Maou, F.; Belloc, A. Scale effect on concrete in tension. In *Materials and Structures*; Springer: Berlin/Heidelberg, Germany, 1994; pp. 437–444.
11. Rastiello, G. Influence de la Fissuration sur le Transfert de Fluides dans les Structures en béton. Stratégies de Modélisation Probabiliste et étude Expérimentale. Ph.D. Thesis, IFSTTAR, Paris, France, 2013.
12. Rita, M.R.; Rossi, P.; Fairbairn, E.d.M.R.; Ribeiro, F.L.B.; Tailhan, J.L.; Andrade, H.C.C.d.; Mota, M.T. Three-Dimensional Probabilistic Semi-Explicit Cracking Model for Concrete Structures. *Appl. Sci.* **2024**, *14*, 2298. [CrossRef]
13. Rossi, P. Fissuration du béton: Du Matériau à la Structure-Application de la Mécanique Lineaire de la Rupture. Ph.D. Thesis, L'Ecole Nationale des Ponts et Chaussées, Paris, France, 1986.
14. Rita, M.R.; Rossi, P.; Fairbairn, E.d.M.R.; Ribeiro, F.L.B. Determination of the Probabilistic Properties of the Critical Fracture Energy of Concrete Integrating Scale Effect Aspects. *Appl. Sci.* **2024**, *14*, 462. [CrossRef]
15. Metcalf, M.; Reid, J.; Cohen, M. *Modern Fortran Explained: Incorporating Fortran 2018*; Oxford University Press: Oxford, UK, 2018. [CrossRef]
16. Curcic, M. *Modern FORTRAN: Building Efficient Parallel Applications*; Manning Publications: Shelter Island, NY, USA, 2021.

17. Silva, A.B.C.G. A Finite Element Thermoregulation Model of The Human Body for Hypothermia Treatment in Adults and Neonates. Ph.D. Thesis, UFRJ, Rio de Janeiro, Brazil, 2016.
18. Weibull, W. *A Statistical Theory of the Strengh of Materials*; Royal Swedish Institute for Engineering Research: Stockholm, Sweden, 1939.
19. Weibull, W. A statistical distribution function of wide applicability. *J. Appl. Mech.* **1951**, *18*, 293–297. [CrossRef]
20. Hahn, G.J.; Shapiro, S.S. *Statistical Models in Engineering*; Wiley Classics Library: New York, NY, USA, 1994.
21. Bazant, Z.; Oh, B. Crack band theory for fracture of concrete. *Mater. Struct.* **1983**, *3*, 155–177.
22. De Borst, R.; Nauta, P. Nonorthogonal cracks in a smeared finite element model. *Eng. Comput.* **1985**, *2*, 35–46. [CrossRef]
23. Bazant, Z.P. Mechanics of distributed cracking. *Appl. Mech. Rev.* **1986**, *39*, 675–705. [CrossRef]
24. Pijaudier-Cabot, G.; Bažant, Z. Nonlocal damage theory. *J. Eng. Mech.* **1987**, *113*, 1512–1533. [CrossRef]
25. Bažant, Z.P.; Pijaudier-Cabot, G. Measurement of Characteristic Length of Nonlocal Continuum. *J. Eng. Mech.* **1989**, *115*, 755–767. [CrossRef]
26. Pijaudier-Cabot, G.; Mazars, J.; Pulikowski, J. Steel-Concrete Bond Analysis with Nonlocal Continuous Damage. *J. Struct. Eng.* **1991**, *117*, 862–882. [CrossRef]
27. Fichant, S.; La Borderie, C.; Pijaudier-Cabot, G. Isotropic and anisotropic descriptions of damage in concrete structures. *Mech. Cohesive Frict. Mater.* **1999**, *4*, 339–359. [CrossRef]
28. Zhou, X.; Feng, B. A smeared-crack-based field-enriched finite element method for simulating cracking in quasi-brittle materials. *Theor. Appl. Fract. Mech.* **2023**, *124*, 103817. [CrossRef]
29. Chapman, B.; Jost, G.; Pas, R.V.D. *Using OpenMP*; The MIT Press: Cambridge, MA, USA, 2007.
30. Pas, R.V.D.; Stotzer, E.; Terboven, C. *Using OpenMP—The Next Step*; The MIT Press: Cambridge, MA, USA, 2017.
31. Mattson, T.G.; He, Y.H.; Koniges, A.E. *The OpenMP Common Core*; The MIT Press: Cambridge, MA, USA, 2019.
32. Chen, H.C.; Lim, A.; Warsi, N.A. An Efficient Multilevel Master-slave Model For Distributed Parallel Computation. In *WIT Transactions on Modelling and Simulation*; WIT Press: Southampton, UK, 1999.
33. Jacinto, D.S. Escalonamento de Aplicações Paralelas: De Clusters Para Grids. Master's Thesis, UFSCar, São Carlos, Brazil, 2007.
34. Coutinho, D.A.M. Implementação Paralela Escalável e Eficiente do Algoritmo Simplex Padrão na Arquitetura Multicore. Ph.D. Thesis, UFRN, Rio Grande do Norte, Brazil, 2013.
35. Stallings, W. *Computer Organization and Architecture*; Pearson: London, UK, 2021.

buildings

MDPI

Article

Study on Seasonal Permafrost Roadbed Deformation Based on Water–Heat Coupling Characteristics

Bo Lu [1], Wen Zhao [1,*], Shengang Li [1], Manman Dong [2], Zhikang Xia [3] and Yunfang Shi [1]

1 School of Resources and Civil Engineering, Northeastern University, Shenyang 110819, China; luboneu@163.com (B.L.); lsglili@163.com (S.L.); 2171109@stu.neu.edu.cn (Y.S.)
2 Department of Engineering Management, Changshu Institute of Technology, Changshu 215500, China; manmandong@cug.edu.cn
3 The Construction Engineering Company of City Group, Hefei 230031, China; xiazhikang518@163.com
* Correspondence: zhaowen@mail.neu.edu.cn

Abstract: The deformation and damage to seasonal permafrost roadbeds, as seasons shift, stems from the intricate interplay of temperature, moisture, and stress fields. Fundamentally, the frost heave and thaw-induced settlement of soil represent a multi-physics coupling phenomenon, where various physical processes interact and influence each other. In this investigation, a comprehensive co-coupling numerical simulation of both the temperature and moisture fields was successfully executed, utilizing the secondary development module within the finite element software, COMSOL Multiphysics 6.0. This simulation inverted the classical freezing–thawing experiment involving a soil column under constant temperature conditions, yielding simulation results that were in excellent agreement with the experimental outcomes, with an error of no more than 10%. Accordingly, the temperature, ice content, and liquid water content distributions within the seasonal permafrost region were derived. These parameters were then incorporated into the stress field analysis to explore the intricate coupling between the moisture and temperature fields with the displacement field. Subsequently, the frost heave and thaw settlement deformations of the roadbed were calculated, accounting for seasonal variations, thereby gaining insights into their dynamic behavior. The research results show that during the process of freezing and thawing, water migrates from the frozen zone towards the unfrozen zone, with the maximum migration amount reaching 20% of the water content, culminating in its accumulation at the interface separating the two. Following multiple freeze–thaw cycles, this study reveals that the maximum extent of freezing within the roadbed reaches 2.5 m, while the road shoulder experiences a maximum freezing depth of 2 m. A continuous trend of heightened frost heave and thaw settlement deformation of the roadbed is observed in response to temperature fluctuations, leading to the uneven deformation of the road surface. Specifically, the maximum frost heave measured was 51 mm, while the maximum thaw settlement amounted to 13 mm.

Keywords: seasonal permafrost roadbed; water–heat coupling characteristics; freeze–thaw cycles; frost heave and thaw-induced settlement

1. Introduction

In recent years, the exponential expansion of rail transit engineering has posed novel challenges to construction endeavors in extreme cold regions, notably in highway construction [1,2]. Permafrost, a unique geological blend of soil and ice, exhibits intricate instabilities, governed by intricate interactions between water, heat, and mechanical forces. Consequently, roadbeds constructed in permafrost regions are susceptible to two primary types of detrimental effects: frost heaving, which results from water expansion upon freezing, and thaw settlement, occurring as the ice melts and the ground subsides [3,4]. It is evident that the physical and mechanical properties of seasonal permafrost roadbeds are predominantly influenced by temperature fluctuations and water content, both of which play pivotal roles in determining their stability and durability [5].

Under fluctuating temperatures, roadbeds within seasonal permafrost zones experience cycles of freezing and thawing, where soil freezing is triggered by the migration of water under temperature gradients. This migration process is intricately intertwined with the phase change between ice and water, where the heat liberated during this transformation decelerates the rate of water migration. Furthermore, the formation of ice acts as a barrier, impeding the movement of unfrozen water. Consequently, the intricate water–heat coupling mechanisms within permafrost represent a critical area of research focus [6,7]. Through field observations and experiments, Tabe established that water migration is the primary driver of frost heaving, introducing the capillary theory [8]. Everett further elaborated on the thermodynamics behind the formation of "ice lenses" in porous materials, considering the damage to roads and buildings caused by freezing, refining the capillary theory [9]. While this theory offers a clear understanding of ice formation mechanisms, its quantitative application remains challenging. In response, Harlan drew parallels between fluid transfer in partially frozen soil and, in unsaturated soil, disregarded ice lens formation to model the effects of temperature and water content on fluid movement [10]. Taylor and Luthin then introduced the concept of "ice impedance", positing that ice hinders water migration in permafrost, thereby simplifying Harlan's model [11].

In recent years, numerous academics have advanced water–heat coupling models that encompass comprehensive frameworks for water–heat migration equations, ice–water phase transitions, and intricate soil deformation calculations. Zhan et al. constructed a water–heat–stress coupling model to investigate the frost heave deformation characteristics of soil slopes in seasonal areas [12]. Booshehrian et al. defined the relationship between unfrozen water content and cryogenic suction, based on the similarity between soil freezing and soil–water characteristic curves for unsaturated soils [13]. Mihara et al. modified the Soil and Water Assessment Tool (SWAT), by incorporating the dynamic change in soil permeability based on the degree of soil freezing [14]. Yin et al. established a mathematical model for the coupling of liquid, vapor, and heat fields [15]. Li et al. established a finite element model, which considered the effect of thermo–hydro–mechanical coupling, to investigate the freezing damage of berms under the influence of the reservoir level and the water migration of dam filling [16]. Lu et al. proposed different modeling approaches by regarding the environmental factor as a constitutive variable and introducing the 3D fractional plastic flow rule into the characteristic stress space [17,18]. Zhang et al. carried out an extensive series of pore water pressure tests under sub-zero conditions, delving into the dynamics of how pore water pressure fluctuates and subsequently influences deformation patterns [19].

Despite the fact that extensive theoretical and experimental investigations have been conducted on permafrost deformation, which have mainly focused on the final results of stable deformation and the main influencing factors, there are limitations to the modeling of moisture transport in permafrost. Scarce attention has been paid to the implications of coupled water–heat variations on the stability of roadbed deformation and, consequently, there exists a notable dearth of models that have been effectively applied to practical engineering contexts.

Hence, this study is based on the construction of the Shuang-Tao expressway roadbed. The theoretical and numerical simulation methods are combined to study the water–heat coupling process and the deformation of the roadbed. First, the coupling model was simplified by incorporating a solid–liquid ratio and relative saturation, considering the influence of ice–water phase transitions on the water–heat distribution. This refined model is then implemented within the COMSOL software framework, to replicate and analyze classic freezing and thawing experiments on soil columns. Through secondary development, the rationality and accuracy of the established model were validated. Subsequently, the validated coupling model was applied to the Shuang-Tao expressway roadbed, leveraging numerical simulations to comprehensively characterize the roadbed's water, heat, and stress states, as well as their dynamic variations over time. Finally, the intricate patterns and trends governing the evolution of the roadbed's water content field, temperature field, and deformation field were investigated, thereby providing valuable insights into the roadbed's performance and durability.

2. The Water–Heat Coupling Model

Seasonal permafrost undergoes cyclic variations driven by external temperature fluctuations, leading to seasonal cycles of freezing and thawing. These processes can induce detrimental effects on roadbeds, including cracking, frost heave, and thaw-induced settlement. The primary underlying mechanism is the migration and phase transformation of water within the roadbed soil, during these temperature-driven transitions. Hence, the present study delves into the alterations in temperature and moisture distributions within roadbeds during freeze–thaw cycles. Leveraging a water–heat coupling mathematical framework, this research employs the COMSOL finite element software to perform comprehensive, coupled simulations of both temperature and moisture fields, achieved through tailored secondary development.

This study assumes the following premises: (1) the roadbed soil is modeled as a homogeneous, isotropic, and porous elastic medium; (2) water migration between frozen and unfrozen zones in the roadbed occurs exclusively in the liquid phase; (3) the soil particles, ice, and water are considered incompressible; (4) the movement of liquid water within the roadbed adheres to Darcy's law; and (5) heat variations in the roadbed soil are primarily attributed to heat conduction and the phase transformations of ice and water, while other modes of heat transfer are deemed negligible.

2.1. Water–Heat Coupling Differential Equations

Due to the elongated nature of the roadbed along the longitudinal axis and its significantly reduced dimensions in perpendicular directions, the analysis of its moisture field can be simplified and focused on the cross-sectional plane, thereby transforming the study into a two-dimensional problem of academic rigor. Based on Darcy's law [20] and the principle of mass conservation [21], the governing equation for the migration of permafrost moisture can be obtained as:

$$\frac{\partial \theta_{\mathrm{u}}}{\partial t} = \frac{\partial}{\partial x}\left[D(\theta_{\mathrm{u}})\frac{\partial \theta_{\mathrm{u}}}{\partial x}\right] + \frac{\partial}{\partial y}\left[D(\theta_{\mathrm{u}})\frac{\partial \theta_{\mathrm{u}}}{\partial y}\right] + \frac{\partial}{\partial y}[K(\theta_{\mathrm{u}})] - \frac{\rho_I}{\rho_w}\cdot\frac{\partial \theta_I}{\partial t} \tag{1}$$

The seasonal fluctuations in moisture within permafrost roadbeds elicit phase transformations, resulting in non-steady-state thermal variations. To tackle this phase change challenge utilizing the specific heat capacity approach, it is assumed that there are no extraneous heat sources within the soil. Drawing upon heat conduction theory, a two-dimensional, non-steady-state temperature field heat conduction differential equation is obtained as Equation (2), which is tailored specifically for roadbeds undergoing phase transformations [22].

$$\rho C_p \frac{\partial T}{\partial t} = \frac{\partial}{\partial x}\left(\lambda\frac{\partial T}{\partial x}\right) + \frac{\partial}{\partial y}\left(\lambda\frac{\partial T}{\partial y}\right) + \rho_I L \frac{\partial \theta_I}{\partial t} \tag{2}$$

Equations (1) and (2) reveal three unknowns, namely the liquid water content, ice content, and temperature, yet only two fundamental equations are provided. To ensure a comprehensive and solvable water–heat coupling system, a third, supplementary relationship is essential. This research introduces the concept of the "solid–liquid ratio", B_1, which quantifies the proportion of solid ice to liquid water within the permafrost, thereby completing the mathematical framework.

$$B_1 = \frac{\theta_I}{\theta_u} = \begin{cases} 1.1\left(\frac{T}{T_f}\right)^B - 1.1 & \left(T < T_f\right) \\ 0 & \left(T > T_f\right) \end{cases} \tag{3}$$

2.2. Water–Heat Coupling Numerical Model

Based on the aforementioned model, this study utilized the secondary development of the PDE module within the COMSOL software to achieve the coupling of the moisture and temperature field. The pertinent equations and boundary constraints are outlined as follows:

$$e\frac{\partial^2 u}{\partial t^2}+d\frac{\partial u}{\partial t}+\nabla\cdot(-c\nabla u-\alpha u+\gamma)+\beta\cdot\nabla u+au=f \text{ in}\Omega \tag{4}$$

$$\mathbf{n}\cdot(c\nabla u+\alpha u-\gamma)+qu=g-h^{\mathrm{T}}\mu \text{ on}\Gamma \tag{5}$$

$$u=r \text{ on}\Gamma \tag{6}$$

By converting Equations (1) and (2) into the coefficient form of the partial differential equation group provided by COMSOL, the coupling solution can be achieved.

(1) Modeling of the water field

Firstly, the partial differential Equation (4) is written in a customized form as:

$$e\frac{\partial^2 S}{\partial t^2}+d\frac{\partial S}{\partial t}+\nabla\cdot(-c\nabla S-\alpha S+\gamma)+\beta\cdot\nabla S+aS=f \tag{7}$$

Next, Equation (1) is transformed into the form of a coefficient-type partial differential equation, as follows:

$$\frac{\partial\theta_u}{\partial t}+\frac{\rho_I}{\rho_w}\frac{\partial\theta_I}{\partial t}=\nabla\left[D(\theta_u)\nabla\theta_u+K_y(\theta_u)\right] \tag{8}$$

Based on the VG hysteresis model and the Gardner permeability coefficient model [23], the saturation degree S of the permafrost can be defined as:

$$S=\frac{\theta_u-\theta_r}{\theta_s-\theta_r} \tag{9}$$

Substituting Equations (3) and (9) into Equation (8) yields:

$$\begin{aligned}&\tfrac{\partial S}{\partial t}\left(\theta_s-\theta_r+(\theta_s-\theta_r)\tfrac{\rho_I}{\rho_w}B_1(T)\right)+(\theta_s-\theta_r)\tfrac{\rho_I}{\rho_w}\tfrac{\partial B_1(T)}{\partial t}\cdot S+\\&\theta_r\tfrac{\rho_I}{\rho_w}B_1(T)=\nabla[D(S)\nabla S+K(S)]\end{aligned} \tag{10}$$

Subsequently, the coefficients of the partial differential equation can be obtained as:

$$\begin{cases}d=\theta_s-\theta_r+(\theta_s-\theta_r)\frac{\rho_I}{\rho_w}B_1(T)\\ c=D(S)\\ \gamma=(0,-K_y(S))\\ a=(\theta_s-\theta_r)\frac{\rho_I}{\rho_w}\frac{\partial B_1(T)}{\partial t}\\ f=-\theta_r\frac{\rho_I}{\rho_w}B_1(T)\end{cases} \tag{11}$$

where e, α, and β are all set to 0.

(2) Modeling of the temperature field

First, the partial differential Equation (4) is rewritten as follows:

$$e\frac{\partial^2 T}{\partial t^2}+d\frac{\partial T}{\partial t}+\nabla\cdot(-c\nabla T-\alpha T+\gamma)+\beta\cdot\nabla T+aT=f \tag{12}$$

Subsequently, Equation (2) is expressed in the form of a coefficient-type partial differential equation, as follows:

$$\rho C_p \frac{\partial T}{\partial t} = \nabla \cdot (\lambda \nabla T) + \rho_I L \frac{\partial \theta_I}{\partial t} \tag{13}$$

Based on the principle of establishing the moisture field model, the following can be derived as:

$$\rho C_p \frac{\partial T}{\partial t} = \nabla \cdot (\lambda \nabla T) + \rho_I L\left((\theta_s - \theta_r) \cdot \left(\frac{\partial B_1(T)}{\partial t} \cdot S + B_1(T) \cdot \frac{\partial S}{\partial t}\right) + \frac{\partial B_1(T)}{\partial t} \cdot \theta_r\right) \tag{14}$$

Consequently, the coefficients of the partial differential equation are obtained as follows:

$$\begin{cases} d = \rho C_p \\ c = \lambda \\ f = \rho_I L\left((\theta_s - \theta_r) \cdot \left(\frac{\partial B_1(T)}{\partial t} \cdot S + B_1(T) \cdot \frac{\partial S}{\partial t}\right) + \frac{\partial B_1(T)}{\partial t} \cdot \theta_r\right) \end{cases} \tag{15}$$

where e, α, λ, a, and β are all set to 0.

2.3. Validation of Numerical Model

To validate the efficacy and practical applicability of the aforementioned heat–moisture coupled model, the model was used to invert an enclosed columnar soil freezing and thawing experiment, as conducted by Xu et al. [24].

2.3.1. Freezing Simulation Validation and Analysis

The test soil sample was composed of silty soil, with a diameter (*d*) of 10 cm and a height (*h*) of 15 cm. The initial moisture content was 18.6%. The top plate temperature was maintained at 0.9 °C, while the bottom plate temperature was set to −2.1 °C. The initial temperature was kept constant at 0.9 °C. The lateral sides of the soil column were insulated and impermeable, and there was no water supply at the bottom boundary. The values of the soil parameters required for the model calculations are listed in Table 1.

Table 1. Table of calculation parameters of soil column.

Parameters	Values	Unit	Definition
a_0	2.59	1	Model parameter
m	0.26	1	Model parameter
l	0.5	1	Ontogenetic coefficient
k_s	10-7	m/s	Permeability coefficient
θ_s	0.42	1	Saturated water content
θ_r	0.02	1	Residual water content
ρ_I	900	kg/m^3	Ice density
ρ_w	1000	kg/m^3	Water density
ρ	2060	kg/m^3	Soil density
L	334.5	kJ/kg	Latent heat of phase change in ice and water
λ_s	1.3	W/(m·K)	Thermal conductivity of the soil
C_{af}	900	J/(kg·K)	Specific heat of the soil

The numerical simulation results of the freezing test were compared with the corresponding test results, as presented in Figure 1. Freezing was initiated from the bottom and resulted in a significant increase in the moisture content in the lower portion compared to the pre-freezing state, while the upper portion exhibited a noticeable decrease. This indicated that water migrated from the top to the bottom during the freezing process. The liquid water content in the freezing zone decreased, leading to an increase in suction. Under the capillary action, moisture from the thawing zone in the upper part of the soil column migrated towards the freezing front. Some of it froze near the freezing–thawing interface, while the rest was impeded by the ice, resulting in an increase in the moisture content in the freezing zone, with the appearance of a peak value. When the temperature field approached stability, the final height of the freezing–thawing interface was approximately 7.5 cm and the simulated temperature at the interface was approximately −0.57 °C. Moreover, compared to the temperature field, the moisture field exhibited hysteresis, with moisture migration not immediately reaching equilibrium once the temperature became stable. Instead, it continued to migrate for a certain period of time, ultimately leading to the appearance of a moisture content peak below the freezing–thawing interface. In addition, it can be seen from Figure 1b that the vertical distribution of the moisture content obtained from the numerical simulation was in good agreement with the test results; the average relative error was 1.5%, indicating the numerical simulation accurately reflected the variation characteristics of the water field during the freezing process of frozen soil.

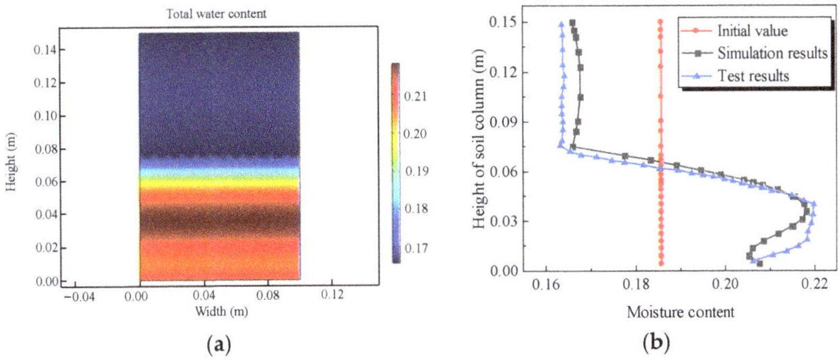

Figure 1. Variation in freezing water content: (**a**) cloud map and (**b**) vertical distribution.

2.3.2. Thawing Simulation Validation and Analysis

The simulation of the thawing test was modeled in reference to the freezing test, with consistent soil and model parameters. However, unlike the freezing test, the initial temperature was kept constant at −2.1 °C, and the water content distribution after 120 h is presented in Figure 2. During the soil column thawing process, the height of the frozen–thawed zone interface and the simulated freezing temperature value were consistent with the freezing test, respectively, 7.5 cm and −0.57 °C. The liquid water content above the frozen–thawed zone interface decreased gradually and uniformly, while the liquid water content below the frozen–thawed zone interface increased vertically from bottom to top. The water content at the frozen–thawed zone interface increased with time and, after the frozen–thawed zone interface stabilized, an increasing trend in the water content could be observed. The variation trend in the total water field in the soil column was basically consistent with the final result of the laboratory thawing test, with an average relative error of 8.5%, verifying the rationality of the water–heat coupling model.

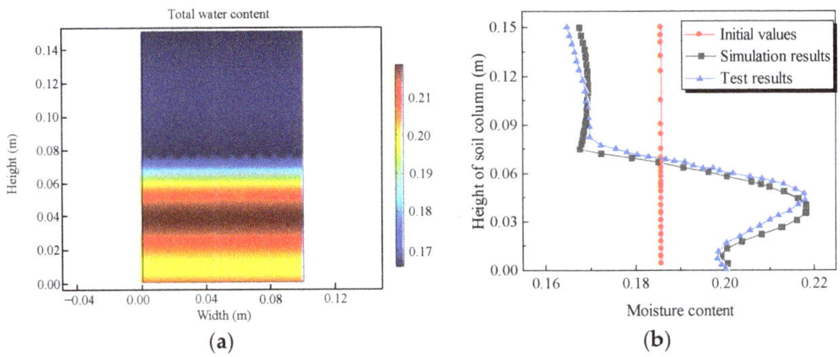

Figure 2. Variation in liquid water content: (**a**) cloud map and (**b**) vertical distribution.

3. Deformation Analysis of Seasonal Frozen Roadbed

The aforementioned numerical simulation of water–heat coupling was applied to investigate the temperature and water fields in the permafrost roadbed of the Shuang-Tao expressway. Firstly, the fully coupled calculation of the water and heat fields of the roadbed was conducted to obtain the distribution of the temperature, ice content, and liquid water content. Then, the data was imported into the stress field to analyze the coupling of the water and temperature fields with the displacement field, and to calculate the deformation and stress of the roadbed, achieving the seasonal permafrost full water–heat coupling and deformation analysis of the roadbed.

3.1. Project Overview

The Shuang-Tao expressway is located in the western region of Jilin province, China, as shown in Figure 3. It is an important longitudinal, regional distribution expressway. The expressway belongs to the mid-temperature zone of the continental monsoon climate, with significant seasonal changes. Spring is dry and windy, summer is hot and rainy, autumn is cool with large temperature differences between day and night, and winter is long and cold.

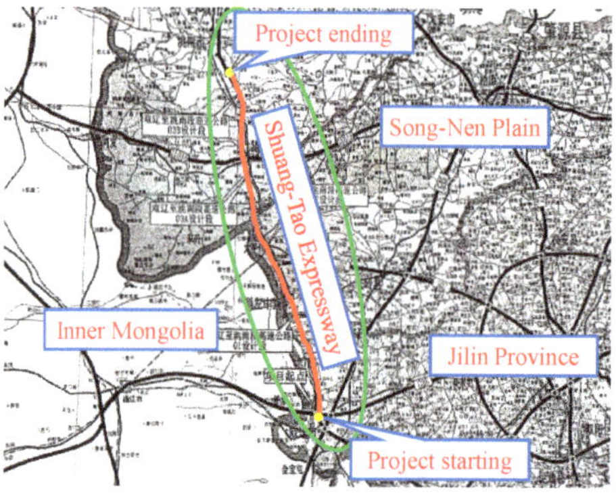

Figure 3. Shuang-Tao expressway project line.

3.2. Water–Heat Coupling Analysis

3.2.1. Parameters of the Model

The width of the roadbed was 27 m, the height was 4 m, and the side slope was 1:1.5. The depth below the roadbed was 10 m. When the boundaries of the model were far enough, the effect of the temperature field of the roadbed on the original seasonal permafrost could be ignored [25]. Therefore, the outward length on each side was taken as 20 m. The roadbed fill was gravel soil and the foundation was powdery clay. The model adopted a free quadrilateral mesh and the roadway section was densified, as shown in Figure 4. The selected parameters for the roadbed are shown in Tables 2 and 3.

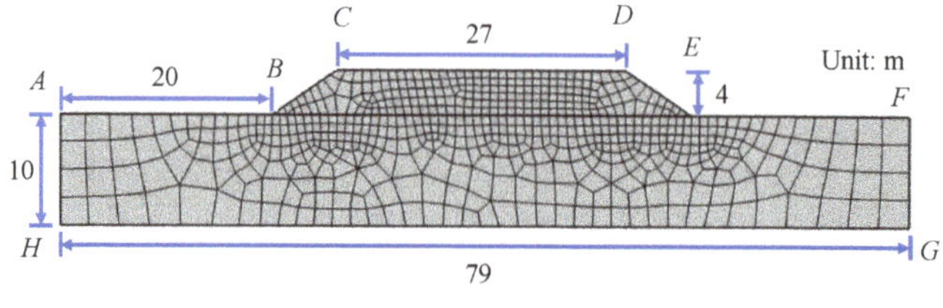

Figure 4. Geometric model and meshing of the roadbed.

Table 2. Thermodynamic parameters.

Materials	Density (kg/m^3)	Thermal Conductivity (W/(m·K))	Specific Heat (J/(kg·K))	B
Gravelly soil	2000	1.13	1210	0.61
Powdery clay	1550	1.125	1241	0.56
Ice	900	2.22	2090	—
Water	1000	0.58	4180	—

Table 3. Hydraulic parameters.

Materials	a_0	m	l	θ_s	θ_r	k_s
Gravelly soil	0.66	0.14	0.5	0.42	0.05	5×10^{-6}
Powdery clay	2.65	0.26	0.5	0.5	0.02	1×10^{-8}

3.2.2. Boundary Conditions and Simulation Process

(1) Temperature boundary conditions

Considering the actual situation, the surface temperature of the roadbed was described by the following equation based on the local temperature:

$$T = T_0 + A \sin\left(\frac{2\pi}{365} \times t + \varphi_0\right) \tag{16}$$

where T is the surface temperature of the roadbed (°C), T_0 is the average annual temperature (°C), A is the annual amplitude of the fluctuation in the daily average temperature (°C), t is time (d), and φ_0 is the first phase.

Based on the temperature in Jilin province, the annual average temperature was taken as 7.4 °C and the annual maximum temperature difference was 52 °C. Taking 18 October 2019, as the starting point for the temperature, φ_0 was taken as π and a vertical temperature gradient was not considered. Hence, the temperature boundary condition of the upper boundary of the roadbed model was as follows:

$$T = 7.4 + 26 \sin\left(\frac{2\pi}{365} \times t + \pi\right) \tag{17}$$

Assuming that the left and right boundaries, AH and FG, respectively, were sufficiently distant, they could be regarded as adiabatic, thereby neglecting any heat exchange across these interfaces. The deeply embedded bottom boundary GH experienced minimal temperature fluctuations due to the external environment, maintaining a nearly constant temperature of 9 °C. The soil's freezing point T_f was set at −0.4 °C.

(2) Moisture boundary conditions

Within the numerical model, the roadbed was considered a closed system, neglecting precipitation and evaporation impacts on moisture. Based on engineering specifications, the initial soil moisture content was set at 12.36%.

(3) Initial conditions

To minimize computational errors, the thermal boundary conditions on the roadbed's top surface were initially derived using Equation (17). Subsequently, the roadbed's temperature field was simulated over 50a, and the stabilized temperature distribution was adopted as the initial condition for subsequent temperature analysis.

3.2.3. Results and Discussion

(1) Temperature field

To investigate seasonal variations in the ground temperature within a permafrost roadbed, a water–heat coupling analysis was conducted, with subsequent analysis of the roadbed temperature distributions every 60 days, commencing from 18 October of the initial year. As presented in Figures 5 and 6, with a decrease in the external temperature, the temperature of the upper boundary of the roadbed dropped to −15 °C on 17 December, reaching the soil's freezing point, resulting in a 1.3 m deep frozen layer within the roadbed. By 15 February of the subsequent year, the upper boundary temperature stabilized at −15 °C, with the frozen depth extending to approximately 2 m, due to the extended freezing duration. Below this depth, temperatures remained above the freezing point, gradually approaching the initial temperature. On 14 April of the same year, with ambient temperatures rising to 7 °C, the roadbed temperature commenced an upward trend. Nevertheless, pockets of sub-zero soil persisted within the roadbed. Heat transfer occurred bidirectionally, with warmth ascending from deep layers and descending from the surface, initiating a thawing process marked by heat absorption. Despite an overall temperature increase, the frozen depth paradoxically expanded from 2 to 2.5 m. On 13 June, with sustained temperature increases, all soil layers within the roadbed surpassed the freezing point, with the temperature nadir observed at a depth of approximately 2.5 m. On 12 August, the roadbed's temperature hovered around 30 °C, with the soil's coolest point located at a depth of approximately 5 m. This phenomenon stems from bidirectional heat transfer, with both shallow and deep, warmer zones simultaneously transferring heat to cooler regions. Consequently, the depth of the lowest temperature shifted from 2.5 to 5 m. In October of the subsequent year, with ambient temperatures declining to approximately 7 °C, the roadbed's temperature distribution mirrored that observed in October of the initial year. Upon crossing the freezing threshold, a fresh freeze–thaw cycle was initiated.

As depicted in Figure 7, the first, second, and third control points were positioned at depths of 1, 2, and 2.5 m, respectively, below the roadbed's centerline. Meanwhile, the fourth, fifth, and sixth control points were situated at depths of 1, 2, and 2.5, respectively, beneath the natural ground surface, 0.5 m lateral to the left slope foot. Figure 8 illustrates the temporal temperature variation curves for the six control points, revealing a sinusoidal trend that aligns with fluctuations in the boundary temperatures. A consistent pattern of temperature variation was evident across all six points. However, the soil's inherent temperature lag response accounts for the delayed increase in temperatures at points 2 and 3 beneath the roadbed, even as point 1 initiates an upward trend. This lag necessitates varying durations for each depth's controlled point to attain maximum or minimum temperatures. During the freezing period, the roadbed soil at equivalent depths exhibited temperatures 1 °C to 2 °C higher than the foundation soil. Similarly, during soil warming, the roadbed soil retained higher temperatures, attributable to its more pronounced temperature variations compared to the foundation soil. Control points 5 and 6 recorded a minimum temperature of approximately 0 °C, confirming a foundation freezing depth of ~2 m, aligned with the observed conditions. Point 2's minimum temperature was also ~0 °C, whereas point 3 dipped below freezing, indicating a roadbed freezing depth of ~2.5 m. This establishes the order of the depth-based influence of the ground temperature as roadbed > natural ground surface.

(2) Moisture field

Based on the time points of the temperature field, the annual variations in liquid water and ice content within the seasonal permafrost roadbed, correlated with the external temperature fluctuations, as depicted in Figures 9–11. On 17 December, the roadbed's upper boundary temperature fell below freezing point, resulting in a 1.3 m deep frozen zone. The ice content within this zone correlated positively with the decreasing temperatures, peaking at 23%, while the liquid water content remained low at 2%. The ice barrier promoted liquid water migration towards the freeze–thaw interface, causing an increase in the liquid water content (max 13%) within the 1.3–2 m zone. On 15 February of the subsequent year, the upper boundary temperature remained low, leading to a downward progression in the freezing front and an expansion of the frozen zone to a 2.5 m depth. Concurrently, a slight increase in the liquid water content was observed proximate to the newly formed freezing surface. By 14 April of this year, the ambient temperature surpassed freezing, initiating thawing in the shallow frozen zone and an increase in the liquid water content. Despite this, pockets of sub-zero temperatures persisted in the roadbed, prompting liquid water migration from both the superficial and deeper layers towards the residual freezing zone, resulting in enhanced liquid water content approximately 1 m deep within this zone. After 13 June, the roadbed underwent complete thawing due to temperature elevation, minimizing water migration. Until the next freeze–thaw cycle, the roadbed's liquid water content distribution remained stable. Water migration led to an increase in the surface water content, from 12.3% to 22%.

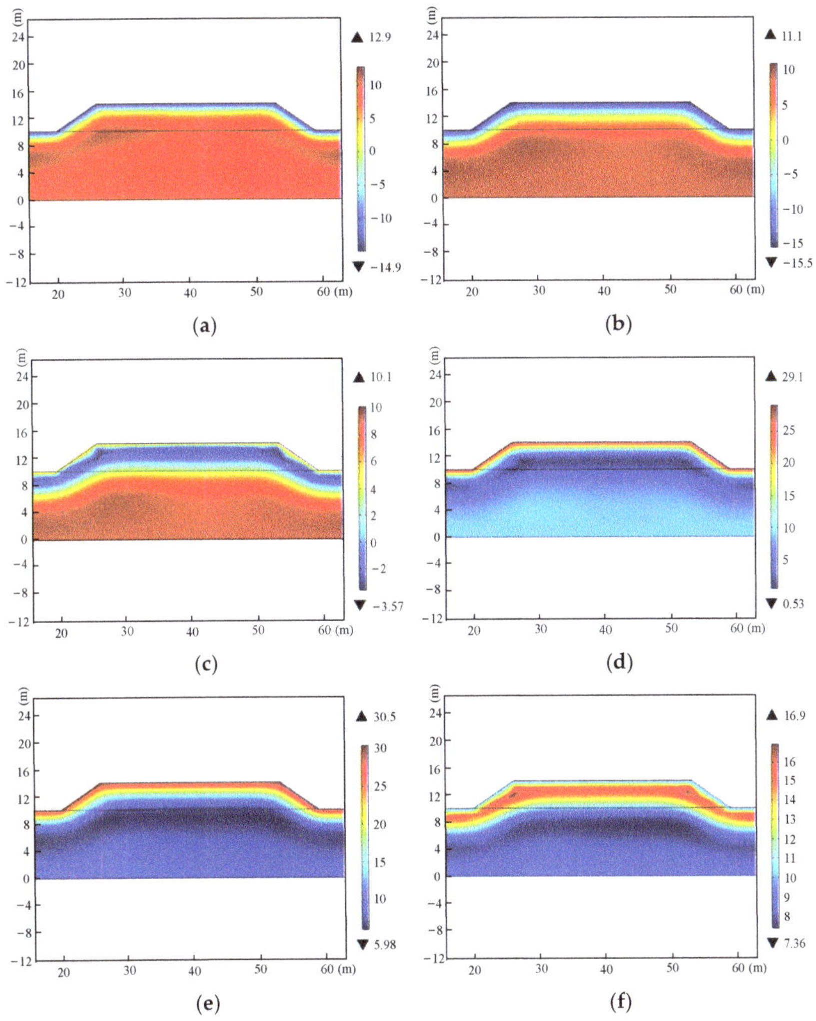

Figure 5. Temperature variation in the roadbed in one year: (**a**) 17 December, (**b**) 15 February, (**c**) 14 April, (**d**) 13 June, (**e**) 12 August, and (**f**) 11 October.

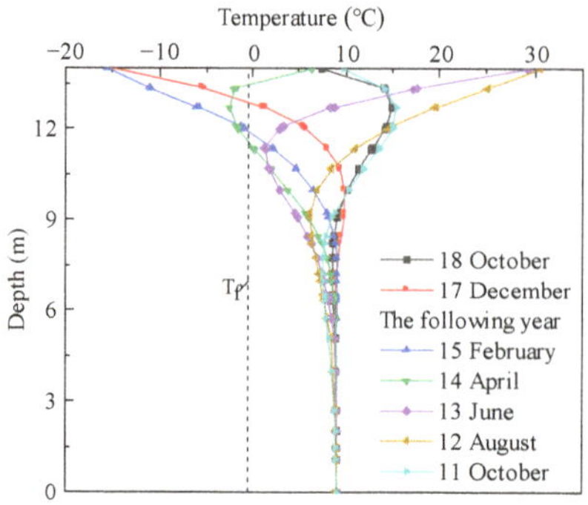

Figure 6. Variation in the temperature in terms of the depth at the center of the roadbed, in one year.

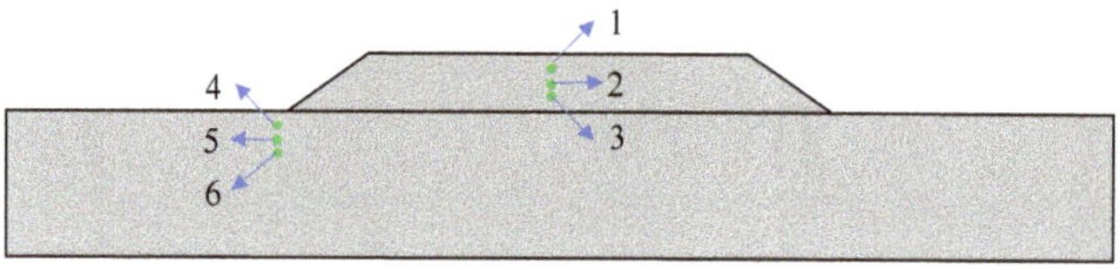

Figure 7. Position of control points in the roadbed.

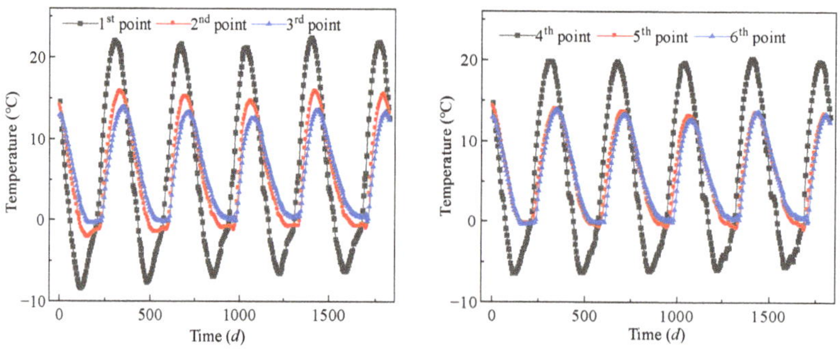

(**a**) (**b**)

Figure 8. Change in control point temperature in 5 years: (**a**) below the roadbed surface and (**b**) below the natural ground surface.

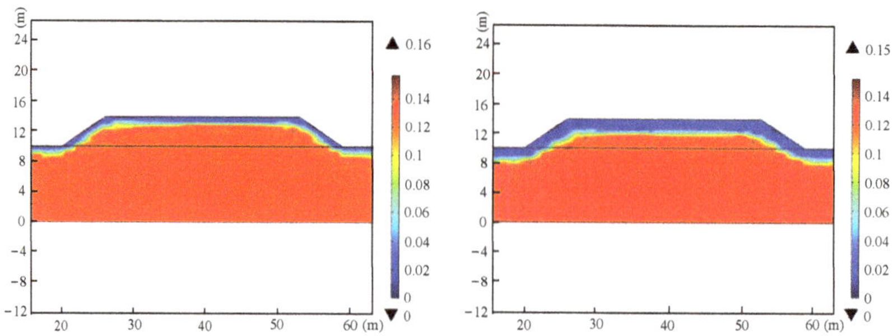

(**a**) (**b**)

Figure 9. *Cont.*

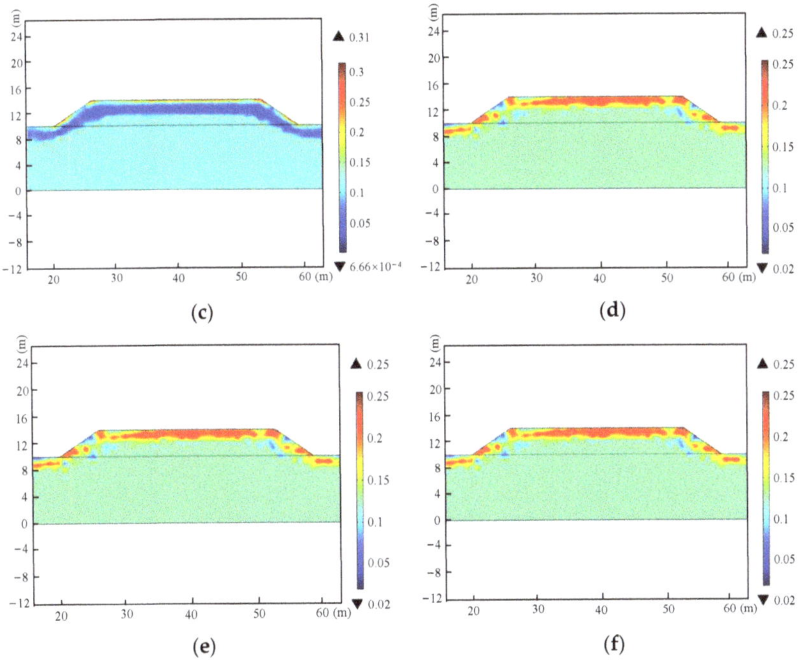

Figure 9. Variation in the liquid water content of the roadbed in one year: (**a**) 17 December, (**b**) 15 February, (**c**) 14 April, (**d**) 13 June, (**e**) 12 August, and (**f**) 11 October.

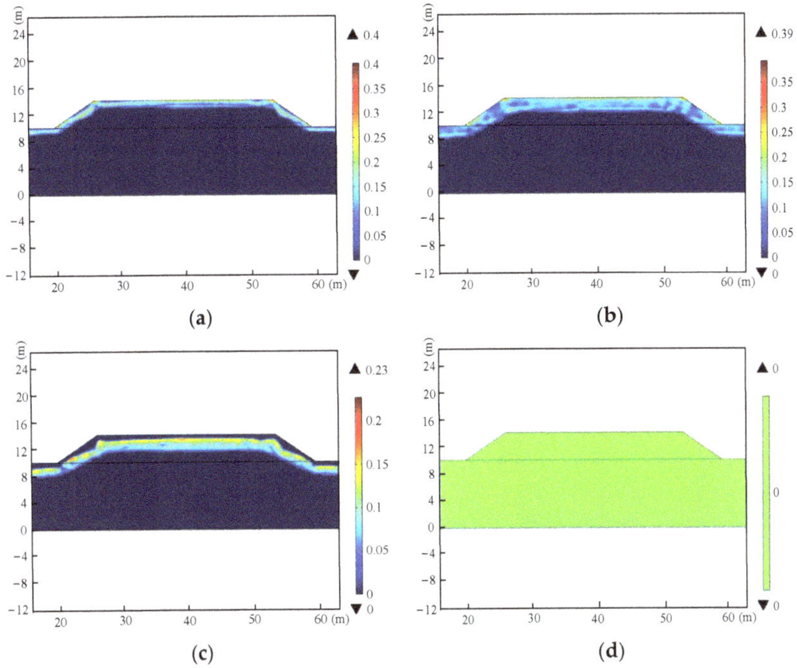

Figure 10. Variation in the ice content of the roadbed in one year: (**a**) 17 December, (**b**) 15 February, (**c**) 14 April, and (**d**) 13 June.

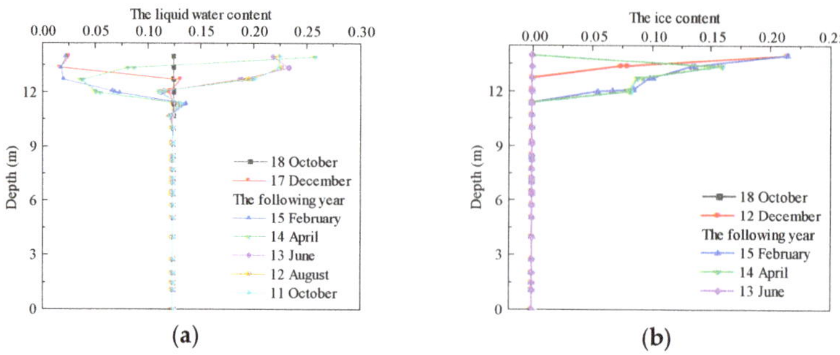

Figure 11. Variations in the liquid water and ice content in the center of the roadbed according to different depths in one year: (**a**) the liquid water content and (**b**) the ice content.

To assess annual variations in the moisture and ice content within the seasonally frozen roadbed, the moisture distributions on August 9th and ice content profiles on January 11th across the second, fourth, sixth, and eighth years post-construction were analyzed, utilizing water–heat coupling simulations, as shown in Figures 12 and 13. The results indicated that over successive freeze–thaw cycles, both the surface and roadbed moisture and ice contents undergo continual redistribution, escalating with the cycle count. After eight years, the roadbed center ice content surged from 0% to 59.2%, while soil below a depth of ~2.5 m remained ice-free, aligning with the roadbed's maximum freezing depth. Furthermore, by the eighth year, the water and ice content near the slope foot peaked at 86% and 106%, respectively, due to its encapsulation by frozen soil and heightened temperature sensitivity, facilitating omnidirectional water migration. The roadbed internal water also converged towards the slope foot under migration effects, leading to cumulative water content (excluding evaporation and drainage). It is inferred that prolonged freeze–thaw cycles will severely compromise the roadbed through frost heave and thaw settlement.

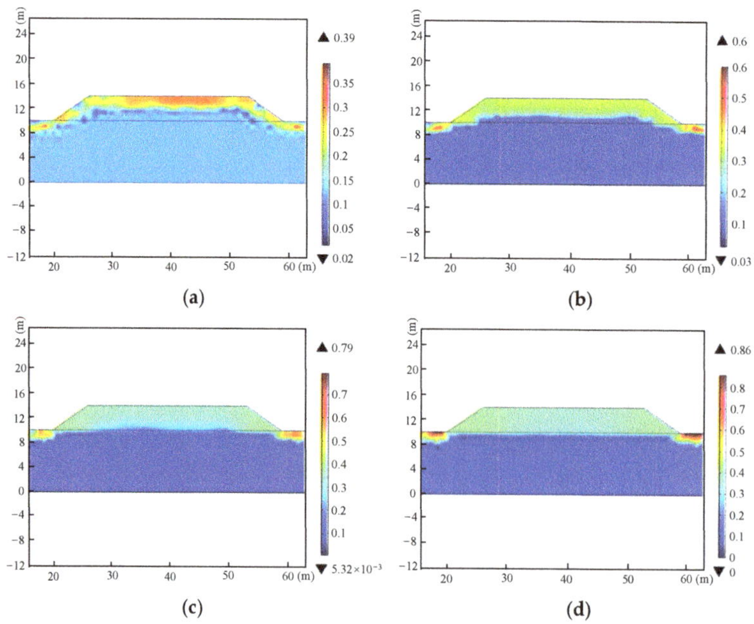

Figure 12. Water content distribution in the roadbed after 8 years: (**a**) after 2 years, (**b**) after 4 years, (**c**) after 6 years, and (**d**) after 8 years.

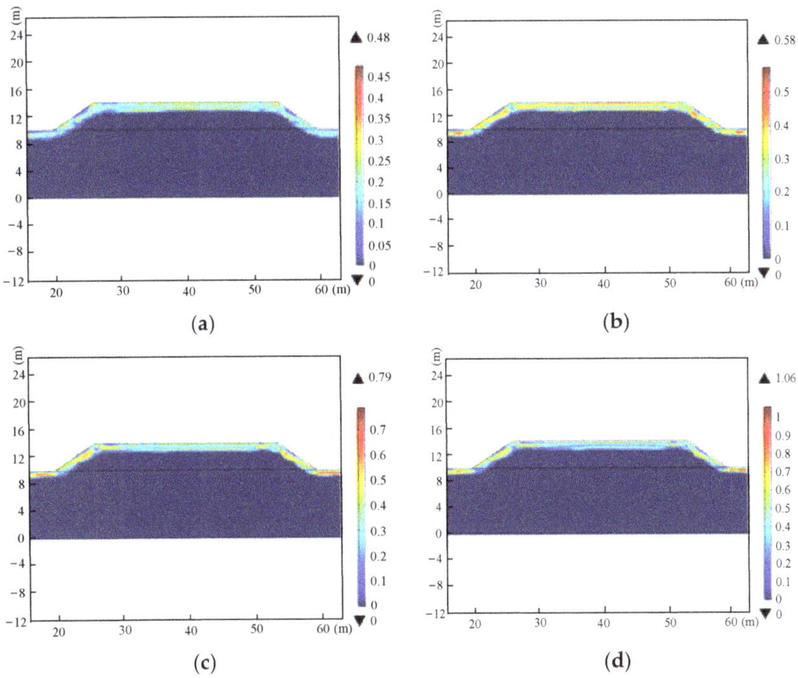

Figure 13. Ice content distribution in the roadbed after 8 years: (a) after 2 years, (b) after 4 years, (c) after 6 years, and (d) after 8 years.

To investigate the water–heat interplay in roadbed soil, time history curves of the liquid water content and temperature, as well as ice content and temperature, were plotted and analyzed, as depicted in Figure 14. The results indicated that the liquid water content at point 1 began to decrease before the temperature reached the freezing point of the soil, and the liquid water at points 1 and 3 slightly increased during the freezing period due to water migration from unfrozen areas. After the temperature reached the freezing point, the ice content rapidly increased, whereas it decreased rapidly as the temperature reached the thawing point. However, the water content field had a slight lag compared to the temperature field. This is because ice is more sensitive to temperature, while liquid water can migrate as the temperature fluctuates.

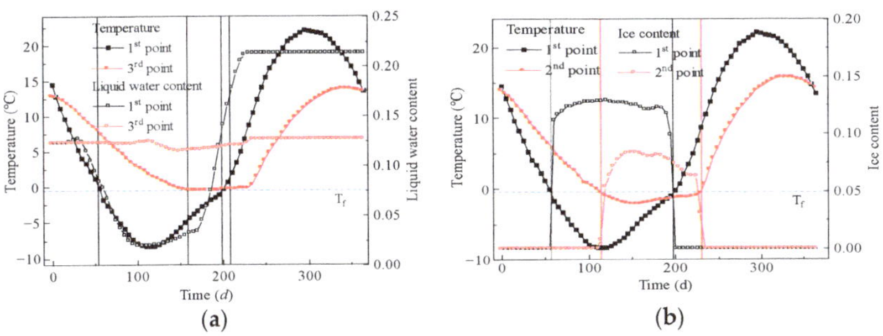

Figure 14. Time history curves: (a) liquid water content and temperature and (b) ice content and temperature.

3.3. Deformation Analysis

3.3.1. Basic Equations for the Stress Field

The stress state of permafrost is complex, including the effects of gravity, external loads, pore water pressure, and frost heave. This investigation neglected the stress redistribution caused by external loads and pore water pressure and only analyzed the stress caused by frost heave. An elastic model was adopted, the soil was analyzed for elastic deformation, disregarding plastic and viscous deformations.

The equilibrium equation is:

$$-\nabla \cdot \sigma = F \tag{18}$$

The geometric equation is:

$$\varepsilon = \nabla u \tag{19}$$

The present constitutive equation is:

$$\{\sigma\} = [c](\{\varepsilon\} - \{\varepsilon_0\}) \tag{20}$$

During the calculation of permafrost strain, it is necessary to consider the phase change of water, the transient strain, and soil strain caused by water migration, which leads to:

$$\varepsilon = \varepsilon^e + \varepsilon_v \tag{21}$$

During the process of water transforming into ice, the volume increases by 9%. The strain ε_v caused by water migration and the freezing phase change is:

$$\varepsilon_v = 0.09(\theta_0 + \Delta\theta - \theta_u) + \Delta\theta + (\theta_0 - n) \tag{22}$$

As liquid water transforms into ice, the volume expands due to freezing and the migration of water from unfrozen zones, alongside in-pore water freezing. Assuming the volume of the soil particles remains constant, the porosity of the soil increases as the volume increases, caused by the water freezing into ice. Hence, frost heave can be quantified through porosity changes, expressed at time t as:

$$e_t = \frac{V_t}{V} \tag{23}$$

Then, the porosity at the time t + dt is:

$$e_{t+dt} = \frac{V_t + dV}{V + dV} \tag{24}$$

The strain during soil freezing is:

$$d\varepsilon_v = \frac{dV}{V} \tag{25}$$

Substituting Equations (23) and (24) into (25) yields:

$$d\varepsilon_v = \frac{\frac{V \cdot e_{t+dt} - V \cdot e_t}{1 - e_{t+dt}}}{V} \approx \frac{de}{1-e} \tag{26}$$

3.3.2. Numerical Modeling

(1) Calculation parameters

The mechanical parameters of the soil for the simulation are shown in Table 4. In the simulation, permafrost was treated as a frozen body composed of soil particles and ice, the elastic modulus of soil and ice were E_s and E_i, respectively, and Poisson's ratio of soil and ice was v_s and v_i, respectively. The content of the soil particles and ice was n_s and n_i, respectively. Thus, the equivalent elastic modulus and Poisson's ratio of permafrost can be obtained as:

$$E = \frac{[n_s E_s(1-2v_i) + n_s E_s(1-2v_s)][n_s E_s(1+v_i) + n_i E_i(1+v_s)]}{n_s E_s(1+v_i)(1-2v_i) + n_1 E_i(1+v_s)(1-2v_s)} \tag{27}$$

$$v = \frac{n_s E_s v_s(1+v_i)(1-2v_i) + n_i E_i v_i(1+v_s)(1-2v_s)}{n_s E_s(1+v_i)(1-2v_i) + n_i E_i(1+v_s)(1-2v_s)} \tag{28}$$

Table 4. Mechanical parameters of the soil.

Materials	Elastic Modulus E (MPa)	Poisson's Ratio v
Gravelly soil	61	0.28
Powdery clay	28	0.3
Ice	-	0.33

The elastic modulus of ice at different temperatures is shown in Table 5. A linear relationship, with a correlation coefficient R^2 of 0.92, can be obtained through linear regression, as shown in Equation (29).

$$E_i = 4363.5 - 323.5T \ (T < T^*) \tag{29}$$

where T is the temperature of the ice and T^* is the freezing point of the soil.

Table 5. Elastic modulus of ice at different temperatures.

Temperature (°C)	−2	−5.4	−8
Elastic modulus (MPa)	4910.0	6342.0	6820.0

(2) Boundary conditions

The roadbed's GH bottom boundary was reinforced with fixed constraints, while its AH and FG side boundaries had roller support. The rest of the roadbed's boundaries were free.

(3) Roadbed freezing and thawing coefficient

The relationship between the freezing coefficient and the ice content was described using an empirical equation, as follows:

$$\eta(x,y) = \begin{cases} A \cdot \omega_I(x,y) - B, & \omega_I(x,y) > C \\ 0 & ,\omega_I(x,y) < C \end{cases} \tag{30}$$

The equation includes empirical coefficients A, B, and C, which can be obtained from Table 6. The mass fraction $\omega_I(x, y)$ of the ice content $\theta_I(x, y)$ is used to express their proportion relationship, as follows:

$$\omega_I(x,y) = \rho_I \cdot \theta_I(x,y)/\rho \tag{31}$$

Table 6. Empirical coefficients.

Soil Types	A	B	C
Gravelly soil	0.0684	0.0028	0.04
Powdery clay	0.08	0.0003	0.003

The thawing coefficient is mainly related to the water content, dry density, and porosity. This investigation used the water content method (ω method) for the calculation [26], and the empirical relationship between the thawing coefficient and water content can be obtained as follow:

$$\eta_0 = K_1(\omega - \omega_c)\% \tag{32}$$

where ω_c represents the initial thawing water content of different soil types, and K_1 is an empirical coefficient related to different soil types. The values of ω_c and K_1 are presented in Table 7.

Table 7. Values of ω_c and K_1 of different soil types.

Soil Types	ω_c (%)	K_1
Gravelly soil	14	0.6
Powdery clay	18	0.7

(4) The deformation analysis steps

First, based on the established water–heat coupling model, the solid mechanics module was added, and the aforementioned water–heat coupling results were imported into the model through interpolation functions. Then, the thermal expansion interface was introduced into the solid mechanics interface, and the relevant parameters and boundary conditions were set. The thermal expansion coefficient was calculated using the secant method. When the temperature fell below the freezing point of the soil, the freezing coefficient was applied. Conversely, when the temperature exceeded the soil's freezing point, the thawing coefficient was used. The thermal expansion coefficients of gravelly soil and powdery clay were converted into IF statement inputs based on Equations (30)–(32). Taking powdery clay as an example, the input expression is as follow:

$$\text{if}(\mathrm{T} < \mathrm{T}_f, Q_1, Q_2) \tag{33}$$

where T_f is the freezing point of powdery clay, and Q_1 and Q_2 define the freezing and thawing coefficient functions of powdery clay. Finally, set the research type as the steady state for model calculations.

3.3.3. Results and Discussion

A study was conducted to investigate the deformation of the road and shoulder during freezing conditions, with displacement–time curves presented in Figure 15. It can be seen that by the 20th day, November's onset saw a persistent decline in ambient temperature, leading to gradual roadbed soil expansion. Notably, the vertical deformation at the right shoulder exceeded that at the roadbed center. This is because the soil on both sides of the roadbed's right shoulder is frozen and a large temperature gradient is formed during the freezing process. It freezes faster than the soil at the center of the roadbed. Therefore, the water in the roadbed soil migrates first to the shoulder, making the freezing and expansion of the soil at the shoulder more obvious. Furthermore, the period spanning from the 20th to 50th day marked a rapid phase of freezing and expansion, with substantial increases in frost heave. Subsequently, the rate of increase diminished on a monthly basis. By the 140th day, the maximum frost heave at the shoulder reached 13 mm, whereas the center of the roadbed experienced a maximum of 7.3 mm. This disparity in deformation between the shoulder and roadbed center will cause uneven deformation of the road.

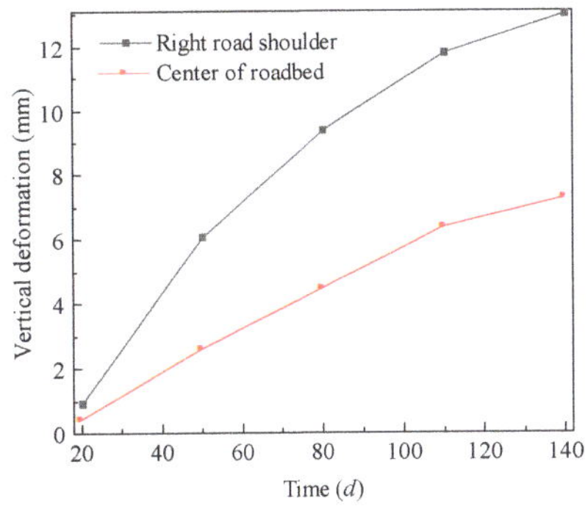

Figure 15. Displacement versus time curve for the middle of the roadbed and the left shoulder during freezing.

An analysis of frost heave deformation on 7 March, annually, was conducted, as depicted in Figure 16, revealing a symmetrical distribution of deformations along the transverse direction. Notably, the roadbed surface exhibited uneven deformation, resulting in a concave shape post-frost heave. Furthermore, the roadbed underwent continuous moisture redistribution due to the influence of water migration, with each freeze–thaw cycle contributing to water accumulation in the upper layers. Neglecting external factors, like drainage and evaporation, this accumulation exacerbated the frost heave deformation over time. By day 140, the shoulder experienced a peak frost heave of 11.7 mm, which increased to 50.5 mm over 7 years. Similarly, the road center attained a maximum frost heave of 7.3 mm on day 140, increasing to 25.2 mm in 7 years. In summary, if the roadbed fill becomes waterlogged, the higher moisture content leads to greater frost heave deformation, and uneven settlement on the roadbed surface will continue to worsen.

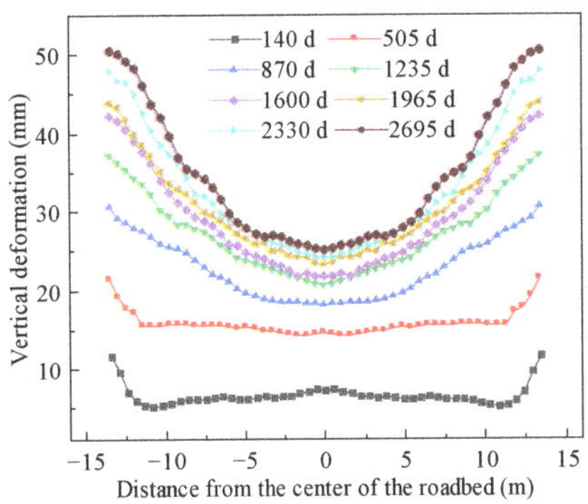

Figure 16. Frost heave deformation of roadbed surface at different times.

An analysis was conducted on the deformation of the roadbed and shoulders during the thawing period, and the displacement versus time curves are shown in Figure 17. It can be seen that after 140 days, the external environment transitioned to a warming phase, causing soil ice to gradually melt as temperatures rose. By the end of May, complete thawing of the roadbed permafrost led to substantial thaw settlement deformation, with the shoulders experiencing the most significant settlement. This is attributed to the preferential heating and thawing of shoulder soils, initiating rapid vertical displacement changes during the initial thaw period. During the thawing process, due to the deep-freezing depth during the freeze-up period, the thawing of the soil at the shoulders took a longer time. Additionally, the water content in the shoulder soil cannot be discharged in a timely manner, causing a larger thaw settlement amplitude, resulting in larger vertical displacement changes at the shoulder. Moreover, between days 140 and 170, there was a rapid phase of thaw settlement development, marked by a steep increase in the settlement growth rate. Subsequently, the thaw settlement rate gradually tapered off on a monthly basis, peaking at 1.2 mm on the 230th day, stabilizing thereafter. This trend stems from ice's sensitivity to temperature; as temperatures rise, the soil ice content rapidly diminishes in the initial stages, leading to a corresponding decrease in frost heave deformation. When the ice completely thawed, water migration within the shallow subgrade layer caused an increase in water migration, leading to thaw settlement deformation. In summary, the difference in frost heave and thaw settlement deformation between the road center and shoulders can cause uneven deformation of the roadbed.

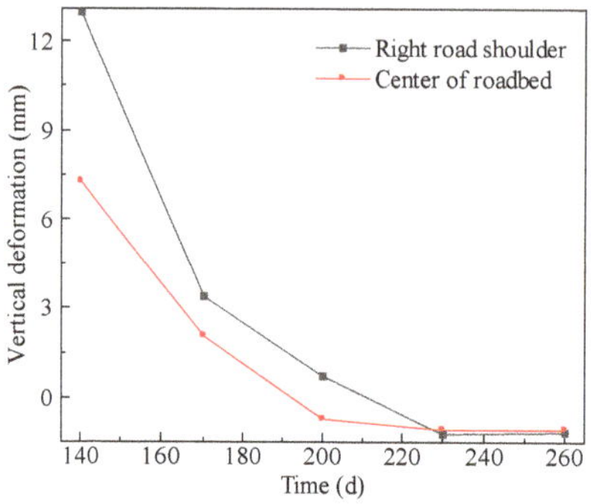

Figure 17. Displacement versus time curve for the middle of the roadbed and the left shoulder during thawing.

The thaw settlement on July 4th of each year was analyzed, as presented in Figure 18. It can be obtained that the thaw settlement deformation of the roadbed was symmetrically distributed along the transverse direction and there was uneven deformation on the roadbed surface, which presented a convex shape after thaw settlement deformation. Additionally, the roadbed underwent constant moisture redistribution due to the influence of water migration. Without considering external factors, such as drainage and evaporation, every freeze–thaw cycle caused water to accumulate on the upper and top surface of the roadbed. This gradual water accumulation enhances thaw settlement deformation over time. The shoulder exhibited a peak thaw settlement of 1.3 mm on the 275th day, escalating to 12.6 mm over 7 years. Comparatively, the road center peaked at 0.76 mm on the 275th day, reaching 9.67 mm after 7 years. Generally, waterlogged subgrade fills experience heightened thaw settlement with rising temperatures due to increased moisture content, exacerbating uneven settlement on the subgrade surface.

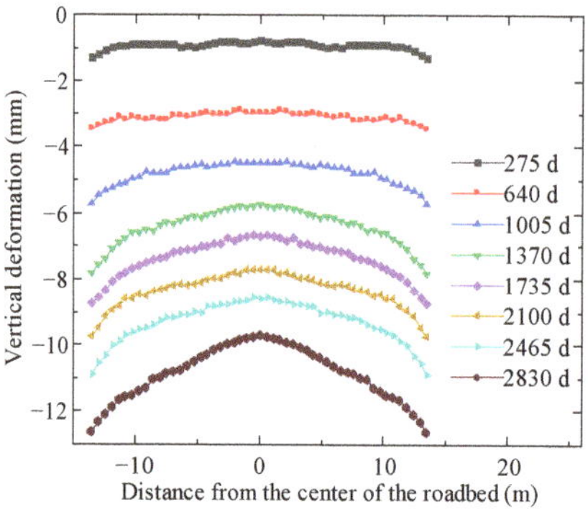

Figure 18. Surface thaw settlement deformation of the roadbed at different times.

3.3.4. Discussion and Limitations

This investigation proposes a water–heat coupling model and implements a computational framework through the secondary development feature of COMSOL Multiphysics, to analyze the influence of water–heat coupling on roadbed deformation. However, this study also has some limitations, as follows:

(1) A sensitivity analysis of the parameters has not been conducted yet. Initially, identify the dominant variables influencing the frost heave/thaw settlement. Subsequently, develop a multi-source data-driven model for evaluating and predicting the safety of seasonal permafrost roadbeds, with soil-sensitive parameters as controlled independent variables, and frost heave/thaw settlement as the target-controlled variables;
(2) This investigation neglected the stress redistribution caused by external loads and pore water pressure and only analyzed the stress caused by frost heave, thus traffic load was not considered. Hence, it is necessary to incorporate dynamic traffic loads to further develop the water–heat coupling computational model under dynamic load conditions.

4. Conclusions

This study presented a combined theoretical and numerical investigation on the water–heat coupling characteristics of a seasonal permafrost roadbed. A water–heat coupling model was established and verified using the classic freeze–thaw test involving a soil column. Furthermore, an analysis of the frost heave and thaw-induced settlement deformations of the roadbed were conducted, focusing on their variations under temporal and thermal fluctuations. The key findings of this research can be summarized as follows:

(1) A water–heat coupling mathematical model is established, of which the calculation is realized by using the secondary development module in the COMSOL software, thereby achieving the fully coupled numerical simulation of the temperature field and the moisture field. The accuracy of the model is verified by the classic freeze–thaw test involving a soil column. The simulation results are in good agreement with the test results, with an error of no more than 10%;
(2) During the freezing and thawing process of the soil, the temperature distribution along the soil column direction is approximately linear, while the water migrates from the positive temperature zone to the negative temperature zone, with the maximum migration amount reaching 20% of the water content. Moreover, the balance of the water field lags behind that of the temperature field;
(3) The duration of the soil freezing period is 4 months, with the maximum freezing depth of the roadbed being 2.5 m and that of the road shoulder being 2 m. After thawing, the water content in the shallow layer of the roadbed increases compared to the initial water content, leading to a decrease in soil strength;
(4) After experiencing multiple freeze–thaw cycles, significant deformation occurs in the roadbed. In the first year, the vertical frost heave deformation of the roadbed is 6 mm, resulting in a concave-shaped road surface; the vertical thaw settlement deformation of the roadbed reaches 1.2 mm, resulting in a convex-shaped road surface. By the seventh year, the maximum frost heave can reach 51 mm, and the maximum thaw settlement can reach 13 mm.

Author Contributions: Conceptualization, B.L., W.Z. and S.L.; methodology, B.L. and W.Z.; software, B.L. and Z.X.; formal analysis, B.L. and Y.S.; investigation, Z.X. and Y.S.; resources, W.Z. and S.L.; data curation, M.D.; writing—original draft preparation, B.L. and M.D.; writing—review and editing, S.L. All authors have read and agreed to the published version of the manuscript.

Funding: This research was funded by the Fundamental Research Funds for the Central Universities (N180104013, Shengang Li) and the China Scholarship Council (202206080043, Bo Lu).

Data Availability Statement: Data will be made available on request.

Conflicts of Interest: Author Zhikang Xia was employed by the company The Construction Engineering Company of City Group. The remaining authors declare that the research was conducted in the absence of any commercial or financial relationships that could be construed as a potential conflict of interest.

References

1. Wu, G.; Xie, Y.; Wei, J.; Yue, X. Freeze-thaw erosion mechanism and preventive actions of highway subgrade soil in an alpine meadow on the Qinghai-Tibet Plateau. *Eng. Fail. Anal.* **2023**, *143*, 106933. [CrossRef]
2. Roh, H.J. Developing a risk assessment model for a highway site during the winter season and quantifying the functional loss in terms of traffic reduction caused by winter hazards conditions. *J. Transp. Eng. Part A Syst.* **2022**, *148*, 04022036. [CrossRef]
3. Cai, H.; Hong, R.; Xu, L.; Wang, C.; Rong, C. Frost heave and thawing settlement of the ground after using a freeze-sealing pipe-roof method in the construction of the Gongbei Tunnel. *Tunn. Undergr. Space Technol.* **2022**, *125*, 104503. [CrossRef]
4. Derk, L.; Unold, F. Effect of temperature gradients on water migration, frost heave and thaw-settlement of a clay during freezing-thaw process. *Exp. Heat Transf.* **2023**, *36*, 585–596. [CrossRef]
5. Huang, X.; Rudolph, D.L.; Glass, B. A Coupled Thermal-Hydraulic-Mechanical Approach to Modeling the Impact of Roadbed Frost Loading on Water Main Failure. *Water Resour. Res.* **2022**, *58*, e2021WR030933. [CrossRef]
6. Zheng, Z.; Bao, Z.; Yang, J.; Cui, M.; Ma, X.; Li, Y. The products transformation and leaching behavior in solidified cement matrices under the coupling actions of "water-heat-chemistry. *Constr. Build. Mater.* **2022**, *323*, 126565. [CrossRef]
7. Sun, W.; Fish, J.; Leng, Z.; Ni, P. PD-FEM chemo-thermo-mechanical coupled model for simulation of early-age cracks in cement-based materials. *Comput. Methods Appl. Mech. Eng.* **2023**, *412*, 116078. [CrossRef]
8. Taber, S. The mechanics of frost heaving. *J. Geol.* **1930**, *38*, 303–317. [CrossRef]
9. Everett, D.H. The thermodynamics of frost damage to porous solids. *Trans. Faraday Soc.* **1961**, *57*, 1541–1551. [CrossRef]
10. Harlan, R.L. Analysis of coupled heat-fluid transport in partially frozen soil. *Water Resour. Res.* **1973**, *9*, 1314–1323. [CrossRef]
11. Taylor, G.S.; Luthin, J.N. A model for coupled heat and moisture transfer during soil freezing. *Can. Geotech. J.* **1978**, *15*, 548–555. [CrossRef]
12. Zhan, Y.; Zhao, M.; Lu, Z.; Liu, G.; Yao, H. Study on water–heat coupling migration law and frost heave effect of soil slope in seasonal frozen regions during groundwater recharge. *Environ. Earth Sci.* **2023**, *82*, 401. [CrossRef]
13. Booshehrian, A.; Wan, R.; Su, X. Hydraulic variations in permafrost due to open-pit mining and climate change: A case study in the Canadian Arctic. *Acta Geotech.* **2020**, *15*, 883–905. [CrossRef]
14. Mihara, K.; Kuramochi, K.; Toma, Y.; Hatano, R. Effect of water movement in frozen soil on long-term hydrological simulation of Soil and Water Assessment Tool in cold climate watershed in Hokkaido, Japan. *Soil Sci. Plant Nutr.* **2024**, *70*, 160–173. [CrossRef]
15. Yin, X.; Liu, E.; Song, B.; Zhang, D. Numerical analysis of coupled liquid water, vapor, stress and heat transport in unsaturated freezing soil. *Cold Reg. Sci. Technol.* **2018**, *155*, 20–28. [CrossRef]
16. Li, Z.; Liu, X.; Sun, Y.; Jiang, X. Numerical simulation of frost heaving damage of earth-rock dam berms in cold regions with thermo-hydro-mechanical coupling. *Cold Reg. Sci. Technol.* **2024**, *223*, 104207. [CrossRef]
17. Lu, D.; Meng, F.; Zhou, X.; Zhuo, Y.; Gao, Z.; Du, X. A dynamic elastoplastic model of concrete based on a modeling method with environmental factors as constitutive variables. *J. Eng. Mech.* **2023**, *149*, 04023102. [CrossRef]
18. Lu, D.; Liang, J.; Du, X.; Ma, C.; Gao, Z. Fractional elastoplastic constitutive model for soils based on a novel 3D fractional plastic flow rule. *Comput. Geotech.* **2019**, *105*, 277–290. [CrossRef]
19. Zhang, H.; Zhang, J.; Zhang, Z.; Zhang, M.; Cao, W. Variation behavior of pore-water pressure in warm frozen soil under load and its relation to deformation. *Acta Geotech.* **2020**, *15*, 603–614. [CrossRef]
20. Deng, Q.; Liu, X.; Zeng, C.; He, X.; Chen, F.; Zhang, S. A freezing-thawing damage characterization method for highway subgrade in seasonally frozen regions based on thermal-hydraulic-mechanical coupling model. *Sensors* **2021**, *21*, 6251. [CrossRef]
21. Botkin, N.D.; Brokate, M.; El Behi-Gornostaeva, E.G. One-phase flow in porous media with hysteresis. *Phys. B Condens. Matter* **2016**, *486*, 183–186. [CrossRef]
22. Biryukov, A.B.; Ivanova, A.A. Diagnosis of temperature conditions in metals when heated in continuous furnaces. *Metallurgist* **2018**, *62*, 331–336. [CrossRef]
23. Tian, K.; Yang, A.; Nie, K.; Zhang, H.; Xu, J.; Wang, X. Experimental study of steady seepage in unsaturated loess soil. *Acta Geotech.* **2020**, *15*, 2681–2689. [CrossRef]
24. Xu, X.; Wang, J.; Zhang, L. *Permafrost Physics*; Science Press: Beijing, China, 2001; pp. 75–98. (In Chinese)
25. Ma, Z.; Lin, C.; Zhao, H.; Yin, K.; Feng, D.; Zhang, F.; Guan, C. Numerical Simulations of Failure Mechanism for Silty Clay Slopes in Seasonally Frozen Ground. *Sustainability* **2024**, *16*, 1623. [CrossRef]
26. Vanmarcke, E. *Random fields: Analysis and Synthesis*; MIT Press: Cambridge, MA, USA, 1983.

buildings

MDPI

Article

A Study on the Early-Stage Mechanical Properties and Uniaxial Compression Constitutive Model of Coral Concrete with Polyoxymethylene Fiber

Jing Wang [1,*], Wenchong Shan [1] and Lipeng Tan [2,3]

1 School of Civil Engineering, Chongqing Industry Polytechnic University, Chongqing 401120, China; danwc@cqipc.edu.cn
2 School of Civil Engineering, Chongqing University, Chongqing 400045, China; 20213883@stu.cqu.edu
3 Key Laboratory of New Technology for Construction of Cities in Mountain Area, Ministry of Education, Chongqing University, Chongqing 400045, China
* Correspondence: wangjing@cqipc.edu.cn

Abstract

To investigate the regulatory mechanism of polyoxymethylene (POM) fiber on the workability and mechanical properties of C30-grade coral aggregate concrete (CAC), this study designed six groups of CAC specimens with varying POM fiber volume fractions (0%, 0.2%, 0.4%, 0.6%, 0.8%, and 1.0%). Cube compressive test, axial compressive test, split tensile test, and flexural tests of CAC specimens after 28 days of curing were conducted, while observing their failure modes under ultimate load and stress–strain curves. The experimental results indicate that POM fiber incorporation significantly reduced the slump and slump flow of the CAC mixtures. The cube compressive strength, axial compressive strength, split tensile strength, and flexural strength of CAC initially increased and then decreased with increasing POM fiber volume fraction, peaking at 0.6% fiber content. Compared to the fiber-free group, these properties improved by 14.78%, 15.50%, 17.01%, 46.13%, and 3.69%, respectively. Analysis of failure modes under ultimate load revealed that POM fibers effectively reduced crack quantity and main crack width, producing a favorable bridging effect that promoted a transition from brittle fracture to ductile failure. However, when fiber volume fraction exceeded 0.8%, fiber agglomeration led to diminished mechanical performance. Based on experimental data, the constitutive relationship established using the Carreira and Chu model achieved a goodness-of-fit exceeding 0.99 for CAC stress–strain curves, effectively predicting mechanical behavior and providing theoretical support for marine engineering applications of coral aggregate concrete. This study provides a theoretical basis for exploiting coral aggregates as low-carbon resources, promoting CAC application in marine engineering, and leveraging POM fibers' reinforcement of CAC to reduce reliance on high-carbon cement. Combined with coral aggregates' local availability (cutting transportation emissions), it offers a technical pathway for marine engineering material preparation.

Keywords: coral aggregate concrete; POM fiber; mechanical properties; constitutive model; low-carbon effect

1. Introduction

With the rapid development of economies and the continuous advancement of infrastructure construction, the demand for building materials continues to rise. Natural aggregates are increasingly depleted in production, leading to resource exhaustion and severe ecological damage. Their extraction and transportation processes also involve substantial energy consumption and carbon emissions. Thus, exploring low-carbon, eco-friendly alternatives to concrete aggregates is essential [1–3]. Recent in-depth studies on coral aggregates, including those by Wang et al. [4] and Sun et al. [5,6], confirm they can be used as concrete aggregates after proper treatment.

Using coral aggregates to replace coarse aggregates in concrete construction has low-carbon and ecological value. It can not only effectively reduce environmental impact, decrease carbon emissions from land sand and gravel mining and transportation, but also significantly reduce project costs and improve the efficiency of coastal engineering projects [7,8]. Due to maritime transportation, the emission of 0.0089 kg of carbon dioxide and the consumption of 0.13 MJ of non-renewable energy are generated for each ton of materials transported for one kilometer. Using local coral waste can avoid the huge energy consumption and carbon dioxide emissions caused by transporting raw materials from the mainland [9]. However, the practical implementation of coral aggregates remains constrained by inherent material limitations such as low compressive strength, porous structure, irregular surface morphology, and higher apparent and bulk densities compared to conventional aggregates. These material properties collectively contribute to operational challenges including heightened water demand during mixing, compromised workability, and difficulties in achieving proper compaction and formwork stability during concrete placement, ultimately affecting the quality and durability of coral-based concrete structures [10–13]. Driven by the low-carbon and environmentally friendly advantages brought by coral aggregates, scholars have conducted research on the modification and enhancement of coral concrete to make up for the shortcomings of coral aggregates. The incorporation of fiber materials has been shown to enhance concrete's crack resistance, tensile strength, and ductility by bridging microcracks and optimizing internal stress distribution [14]. Advances in polymer-based synthetic fiber technology have facilitated the development of high-performance synthetic fibers, providing an effective pathway to improve the mechanical properties of coral aggregate concrete (CAC). Furthermore, experimental studies have demonstrated that synthetic fibers significantly enhance the flexural strength, tensile resistance, and crack inhibition capacity of concrete [15,16]. For example, modified polypropylene fibers can enhance the interfacial bonding performance, fiber pull-out resistance, and fracture toughness of ultra-high performance concrete (UHPC), which in turn reduces the dependence on high cement content and lowers carbon emissions [17]. Polyethylene fibers, by contrast, can improve the interlayer adhesion, impact toughness, and compressive strength of 3D-printed cement-based composites [18], as well as the mechanical properties of fly ash/slag-based geopolymer concrete. Since geopolymers themselves possess low-carbon characteristics, their combination with fibers can further reduce the carbon footprint of the material [19].

Polyoxymethylene (POM) fiber, a novel synthetic polymeric material, exhibits exceptional mechanical properties, long-term durability, and outstanding resistance to chemical corrosion, particularly in alkaline environments and seawater immersion. Remarkably, it maintains homogeneous dispersion stability at high dosages without agglomeration, attributed to its low surface energy and high crystallinity. These synergistic properties have enabled its widespread adoption in marine engineering, fisheries, equipment manufacturing, and precision instrumentation [20].

In the field of concrete performance enhancement, Zhang et al. [21] incorporated POM fibers with varying lengths and volume fractions into notched beams and concrete cubic specimens, conducting three-point bending and splitting tensile tests. The experimental results demonstrated that the increase in fiber length and content effectively enhanced the crack resistance, tensile strength, and fracture toughness of the concrete material. Hua et al. investigated the influence of POM fibers on the bond performance between bimetallic steel bars and seawater sea-sand concrete (SWSSC) through laboratory bond tests. The experimental results revealed that the SWSSC specimens incorporating POM fibers exhibited intact stainless-steel cladding on the bimetallic bars post testing, without cracks, fracture zones, or evidence of galvanic corrosion [22]. Wang et al. [23] systematically documented the apparent morphology and mass loss of POM fiber-reinforced seawater sea-sand concrete (SWSSC) subjected to thermal cycling (heating–cooling processes), and evaluated its compressive and tensile properties. The experimental results demonstrated that SWSSC incorporating POM fibers exhibited enhanced cube compressive strength and splitting tensile strength after exposure to 400 °C, compared to unmodified specimens. Yan et al. [24] investigated the fracture behavior of airport pavement concrete with varying fiber contents. The results demonstrated that increasing fiber content enhanced crack resistance, with fracture toughness increasing by approximately 50% and crack propagation being effectively suppressed when the fiber concentration exceeded 1.1%. Xue et al. [25] studied the effects of POM fibers on the workability, early cracking behavior, and mechanical properties of SWSSC. The results showed that as the fiber volume fraction increased, the workability of SWSSC decreased accordingly. POM fibers significantly improved the early crack resistance of SWSSC. When the POM fiber volume fraction was 0.6%, the cubic compressive strength of SWSSC reached its highest value. When the POM fiber volume fraction was less than 0.6, the axial compressive strength of SWSSC initially increased and then decreased, while it increased when the fiber volume fraction exceeded 0.6. At a volume fraction of 0.6, the splitting tensile strength and flexural strength of SWSSC were also the highest. The above studies found that the length and dosage of POM fibers have certain impacts on the mechanical properties and ductility of concrete, which aligns with conclusions from other researchers [20,26–28]. The current research on fiber concrete mainly focuses on steel/polypropylene fibers combined with ordinary aggregate systems. However, research on the mechanisms by which POM fibers influence the mechanical properties of CAC remains limited. Specifically, studies on the deformation characteristics and reinforcement mechanisms of POM fibers in CAC under various external dosages are scarce. Therefore, in this study, through a 0–1% volume dosage gradient experiment, the strengthening mechanism of POM fibers in coral aggregate concrete (CAC) was systematically revealed, providing important data support for the research on the mechanism of POM fibers affecting the mechanical properties of CAC.

To investigate the early mechanical properties of POM fiber-reinforced CAC specimens, this study designed CAC specimens with gradient POM fiber volume fractions (0%, 0.2%, 0.4%, 0.6%, 0.8%, 1.0%) to systematically evaluate the workability of CAC mixtures and the evolution of their early mechanical behaviors. In accordance with the GB/T 50081-2019 standard [29] for test methods of concrete physical and mechanical properties, tests were conducted on cubic compressive strength, axial compressive strength, splitting tensile strength, and flexural strength. By analyzing the failure modes of the specimens under ultimate load, the synergistic mechanism at the fiber-matrix interface was elucidated. Based on experimental data and theoretical analysis, the optimal range of fiber content was determined. The deformation characteristics of CAC specimens were analyzed using stress–strain curves under different POM fiber contents, and a constitutive model for POM fiber-reinforced CAC was established. As the capacity for marine resource development continues to improve, this provides a theoretical basis for the design and application of CAC in offshore artificial islands, breakwaters, reef ports, and coastal infrastructure projects, and promotes the utilization of coral aggregate as a low-carbon resource. It is conducive to achieving the sustainable development of marine resources and the environment.

2. Property of Material

2.1. Raw Materials

The coral aggregates used in this study were sourced from the South China Sea. The coral debris waste was generated during the construction of marine projects such as ports, docks, and breakwaters. As shown in Figure 1a, the coarse aggregate in the concrete consisted of coral aggregates with a continuous particle size range of 5–16 mm. The mass proportions of particles with diameters between 4.75 and 9.5 mm and 9.5–16 mm were 50% and 50%, respectively. The physical parameters are provided in Table 1.

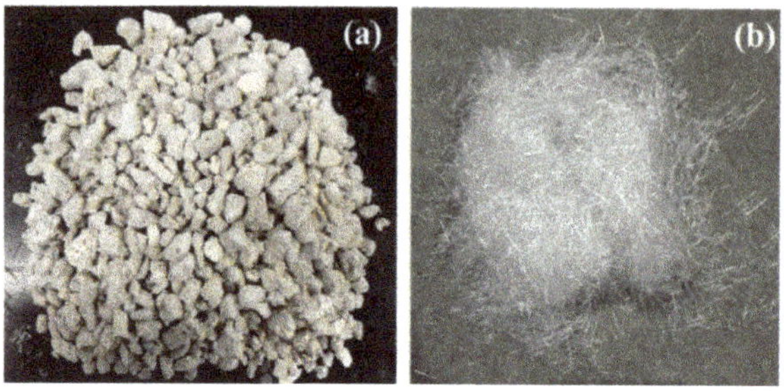

Figure 1. The material of CAC: (**a**) Coral aggregate; (**b**) POM fiber.

Table 1. Physical properties of coral aggregate.

Apparent Density	Bulk Density	Void Content	1 h Water Absorption	Cylindrical Strength
1886 kg/m^3	963 kg/m^3	49.9%	12.1%	2.9 MPa

Note: All indexes listed here are measured referring to GB/T 17431.2-2010 [30] lightweight aggregates and its test methods—part 2: test methods for lightweight aggregates.

The POM fibers used in this study were manufactured by Yunnan Yuntianhua Co., Ltd. (Shuifu City, China). The morphological characteristics of the fibers are illustrated in Figure 1b, with their physical parameters summarized in Table 2. The POM fibers exhibit a diameter of 0.2 mm and a controlled length of 12 mm.

Table 2. Physical properties of POM fiber.

Melting Point	Tensile Strength	Elongation	Elastic Modulus	Density
165 °C	967 MPa	18%	8 GPa	1400 kg/m^3

Note: All indexes listed here are measured referring to GB/T 21120-2018 [31] ynthetic fibers for cement concrete and mortar.

In the CAC specimens, fine aggregates consisted of river sand, with Portland cement Type P·O 42.5 serving as the binder. Freshwater was used as mixing water, and a polycarboxylate superplasticizer (PCE) with a water reduction rate of 35% was incorporated, complying with the requirements of GB 8076-2008 [32] concrete admixtures.

2.2. Mix Proportion and Specimen Preparation

The mixing process for CAC preparation was conducted in accordance with the JGJ/T 12-2019 [33] technical standard for the application of lightweight aggregate concrete. We weighed the coarse aggregate, fine aggregate, and cement according to the mixing ratio, poured them into the mixer, set the speed at 50 rpm, and carried out dry mixing for 1 min.

Then, we added water and water reducer to the mixer and mixed for another 1 min. Finally, we evenly sprinkled POM fibers into the concrete mixture and continued to mix for 2 min. This operation ensured the uniform distribution of POM fibers in the concrete. The detailed experimental design and mass proportions of constituent materials are presented in Table 3, where NF denotes plain CAC without POM fibers and serves as the control group. Six fiber volume fractions (ρ) were investigated: 0%, 0.2%, 0.4%, 0.6%, 0.8%, and 1.0%, with corresponding specimen designations NF, FC0.2, FC0.4, FC0.6, FC0.8, and FC1.0.

Table 3. Mix proportions.

Types	W/B	Cement (kg/m^3)	Coral Aggregate (kg/m^3)	Sand (kg/m^3)	Water (kg/m^3)	POM Fiber (kg/m^3)	Fiber Volume Fraction (%)
F	0.3	500	640	1074	150	0	0
FC-0.2	0.3	500	640	1074	150	2.8	0.2
FC-0.4	0.3	500	640	1074	150	5.6	0.4
FC-0.6	0.3	500	640	1074	150	8.4	0.6
FC-0.8	0.3	500	640	1074	150	11.2	0.8
FC-1.0	0.3	500	640	1074	150	14	1.0

3. Workability

As shown in Figure 2a,b, the slump and slump flow of concrete specimens were measured using a standard slump cone, following the test methods specified in the GB/T 50080-2016 standard [34] for test methods of performance of ordinary concrete mixtures. The slump flow test was performed twice, with the average value adopted as the result. The measured slump and slump flow values are presented in Figure 2c.

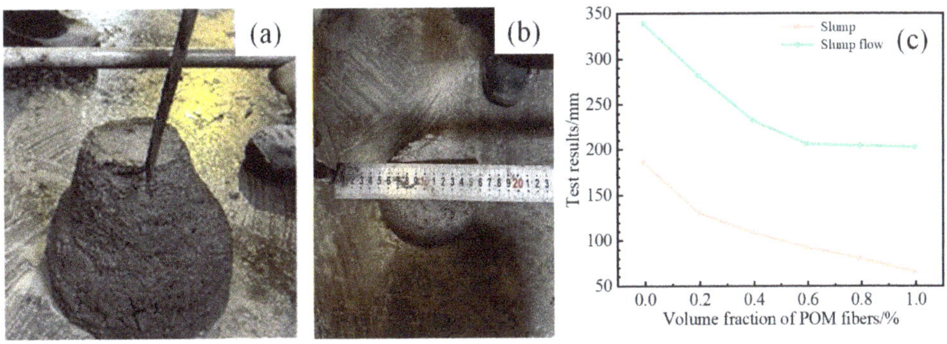

Figure 2. Workability test: (**a**) slump test; (**b**) slump flow test; (**c**) test results of workability of coral concrete.

From Figure 2c, it is evident that both slump and slump flow decrease with increasing POM fiber volume fraction in CAC. Compared to the control group (NF), the CAC specimen FC-1.0 exhibited reductions of 120 mm in slump and 136 mm in slump flow. Due to the addition of fibers, the fluidity of the mixture decreased significantly. Therefore, during engineering applications, the mix ratio needs to be adjusted according to actual requirements, such as using shorter and finer fibers to reduce the demand for the encapsulated slurry and adding silica fume and fly ash to reduce the friction at the fiber–slurry interface.

4. Mechanical Properties

4.1. Mechanical Test Methods and Test Results

According to the GB/T 50081-2019 standard for test methods of physical and mechanical concrete properties, CAC specimens with varying POM fiber volumes were prepared. After the specimens were cured in a standard curing room at 20 ± 2 °C and 95% relative humidity for 28 days, cube compressive strength, axial compressive strength, splitting tensile strength, and flexural strength tests were conducted. As shown in Figure 3a, the tests were performed using an SHT-series computer-controlled electro-hydraulic servo universal testing machine. The experimental procedures for these four strength tests are illustrated in Figure 3b to Figure 3e, respectively. Table 4 summarizes the specimen dimensions and strength calculation formulas for cube compressive strength, axial compressive strength, splitting tensile strength, and flexural strength. Each group consists of three specimens, making a total of 72 specimens. The tests were repeated three times, and the results were taken as the average of the three specimens. The cube compressive strength, splitting tensile strength, flexural strength, and axial compressive strength of CAC specimens with varying POM fiber volume fractions were experimentally evaluated, and the results are summarized in Table 5.

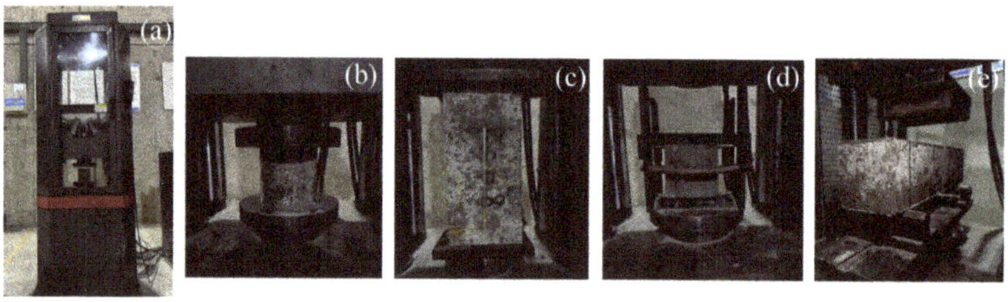

Figure 3. Mechanical Test: (**a**) universal testing machine; (**b**) cube compressive test; (**c**) axial compressive test; (**d**) splitting tensile test; (**e**) flexural test.

Table 4. Specimen dimensions and calculation formula of mechanical test.

Mechanical Test	Specimen Dimensions (mm^3)	Calculation Formula	Parameter Definitions
Cube compressive test	100 × 100 × 100	$f_{cu} = \frac{F}{A}$	f_{cu}: Cube Compressive strength, MPa F: Specimen failure load, N A: Specimen bearing area, mm^2
Axial compressive test	150 × 150 × 300	$f_{cp} = \frac{F}{A}$	f_{cp}: axial compressive strength, MPa
Splitting tensile test	100 × 100 × 100	$f_{ts} = \frac{2F}{\pi A}$	f_{ts}: splitting tensile strength, MPa
Flexural test	100 × 100 × 400	$f_f = \frac{Fl}{bh^2}$	f_f: flexural strength, MPa l: span between supports, mm b: specimen cross-sectional width, mm h: specimen cross-sectional height, mm

Table 5. Mechanical properties of CAC specimens with different POM fiber contents.

Number	f_{cu} (MPa)	f_{cu}-Mean (MPa)	f_{ts} (MPa)	f_{ts}-Mean (MPa)	f_f (MPa)	f_f-Mean (MPa)	f_{cp} (MPa)	f_{cp}-Mean (MPa)
NF-1	35.22		2.05		3.17		30.38	
NF-2	33.61	34.5	1.81	1.94	2.93	2.97	31.45	31.15
NF-3	34.63		1.95		2.81		31.61	
FC-0.2-1	36.32		1.98		3.20		32.58	
FC-0.2-2	35.55	35.9	1.84	1.99	3.68	3.42	32.13	32.69
FC-0.2-3	35.93		2.14		3.39		33.37	
FC-0.4-1	38.31		2.21		4.25		36.14	
FC-0.4-2	37.29	38.5	1.99	2.11	4.16	4.15	34.04	34.53
FC-0.4-3	39.96		2.12		4.05		33.41	
FC-0.6-1	41.11		2.38		4.27		37.26	
FC-0.6-2	39.23	39.6	2.29	2.27	4.71	4.34	34.54	35.98
FC-0.6-3	38.37		2.13		4.03		36.14	
FC-0.8-1	39.14		2.40		4.12		35.65	
FC-0.8-2	38.86	38.6	2.12	2.21	4.64	4.24	34.15	34.59
FC-0.8-3	37.89		2.10		3.95		33.97	
FC-1.0-1	38.20		1.93		3.18		34.17	
FC-1.0-2	37.31	37.0	2.11	2.09	3.90	3.70	32.85	33.40
FC-1.0-3	35.44		2.23		4.01		33.19	

4.2. Descriptive Statistics Analysis of the Mechanical Test Results

The mechanical test results were subsequently analyzed by calculating statistical parameters including standard deviation (SD), coefficient of variation (CV), and 95% confidence intervals to quantify data variability (95% CI).

As shown in Tables 5 and 6, the cube compressive strength of CAC ranges from 34.5 MPa to 39.6 MPa, with standard deviations fluctuating between 0.54 and 1.14. The corresponding coefficients of variation (CV) values fall within 0.0087–0.0311, indicating low data dispersion. The splitting tensile strength of CAC ranges from 1.81 MPa to 2.40 MPa, with standard deviations varying between 0.09 and 0.14, corresponding to CV values of 0.0429–0.0621, demonstrating low variability. The flexural strength of CAC ranges from 2.81 MPa to 4.71 MPa, with SD varying between 0.08 and 0.37, corresponding to CV values of 0.0197–0.0577, demonstrating low variability. The maximum mean flexural strength of CAC reaches 4.34 MPa at a POM fiber volume fraction of 0.6, with a 95% confidence interval of 3.87–4.80 MPa. The axial compressive strength of CAC varies between 30.38 MPa and 37.26 MPa. The standard deviations vary between 0.51 and 1.17, corresponding to CV values of 0.0168–0.0338, indicating low data variability.

Table 6. The statistical analysis of descriptive parameters for mechanical test results.

Number	f_{cu}			f_{ts}			f_f			f_{cp}		
	SD	CV	95% CI	SD	CV	95% CI	SD	CV	95% CI	SD	CV	95% CI
NF	0.67	0.0193	33.4/35.6	0.1	0.0508	1.77/2.10	0.15	0.0504	2.72/3.22	0.55	0.0175	30.2/32.0
FC-0.2	0.31	0.0087	35.4/36.5	0.12	0.0617	1.79/2.19	0.2	0.0577	3.10/3.75	0.51	0.0157	31.8/33.5
FC-0.4	1.1	0.0286	36.7/40.3	0.09	0.0429	1.96/2.26	0.08	0.0197	4.02/4.29	1.17	0.0338	32.6/36.4
FC-0.6	1.14	0.0289	37.7/41.5	0.1	0.0456	2.1/2.44	0.28	0.0649	3.87/4.80	1.12	0.031	34.1/37.8
FC-0.8	0.54	0.0139	37.7/39.5	0.14	0.0621	1.98/2.43	0.29	0.0693	3.75/4.72	0.75	0.0218	33.4/35.8
FC-1.0	1.15	0.0311	35.1/38.9	0.12	0.059	1.89/2.29	0.37	0.0996	3.09/4.30	0.56	0.0168	32.5/34.3

Based on the statistical results in Table 6, all mechanical properties of CAC specimens exhibit low variability. The CV for all measured properties is less than 0.07, and the absolute values of the SD are small with gentle fluctuations. This indicates that the test data have high repeatability and stability and are less affected by random errors. Among them, the flexural strength reaches the maximum value when the POM fiber volume fraction is 0.6, and the 95% confidence interval is reasonable, providing a reliable statistical basis for optimizing the mechanical properties of the material.

4.3. Inferential Statistics of the Mechanical Test Results

To improve the scientific rigor of this study and verify whether the differences in mechanical properties under different POM fiber contents are statistically significant, inferential statistical analyses were performed on the experimental results of axial compressive strength, cube compressive strength, splitting tensile strength, and flexural strength. Both one-way analysis of variance (ANOVA) and Tukey HSD tests, two widely used statistical methods, were conducted.

The general ANOVA model is expressed as Equation (1):

$$Y_{ij} = \mu + \alpha_i + \epsilon_{ij} \tag{1}$$

The F-value is computed as Equation (2):

$$F = \frac{MS_{between}}{MS_{within}} = \frac{SS_{between}/df_{between}}{SS_{within}/df_{within}} \tag{2}$$

where Y_{ij} is the observed value, μ is the overall mean, α_i is the effect of the i-th group, and ϵ_{ij} is the residual error, $MS_{between}$ is the mean square between groups, and MS_{within} is the mean square within groups.

The ANOVA results are shown in Table 7. For the four strength parameters, the p value of f_{ts} was greater than 0.05, which is denoted by '×' in the table, indicating no statistically significant differences among the splitting tensile strengths of CAC specimens with different POM fiber contents. In contrast, the p values of f_{cu}, f_f and f_{cp} were less than 0.05, which is denoted by '√' in the table, confirming the presence of statistically significant differences in cube compressive strength, flexural strength, and axial compressive strength among the specimens.

Table 7. The results of ANOVA and Tukey HSD tests.

Mechanical Property	ANOVA F-Value	ANOVA p-Value	Significance
f_{cu}	9.3918	0.000782	√
f_{ts}	2.4404	0.095400	×
f_f	9.3986	0.000796	√
f_{cp}	8.4737	0.001240	√

The p-value is less than 0.05 in the ANOVA test, so it is necessary to conduct a Tukey HSD test as a post hoc analysis. This ensures a more detailed understanding of the pairwise differences and provides stronger support for the interpretation of experimental results.

For post hoc comparisons, Tukey's HSD was applied, as shown in Equation (3):

$$HSD = q_\alpha \sqrt{\frac{MS_{within}}{n}} \tag{3}$$

where q_α is the studentized range statistic, and n is the sample size per group.

The Tukey HSD test results for the significant mechanical properties are presented in Table 8. The critical HSD values were calculated using Equation (3), with the studentized range statistic $q_{0.05}$= 4.75 for α = 0.05, k = 6 groups, and df = 12.

Table 8. Tukey HSD post hoc test results for significant mechanical properties.

Mechanical Property	HSD Critical Value	Significant Pairwise Comparisons ($p < 0.05$)
f_{cu} (MPa)	2.9633	NF vs. FC-0.4, NF vs. FC-0.6, NF vs. FC-0.8, FC-0.2 vs. FC-0.6
f_f (MPa)	0.8329	NF vs. FC-0.4, NF vs. FC-0.6, NF vs. FC-0.8, FC-0.2 vs. FC-0.6
f_{cp} (MPa)	2.7593	NF vs. FC-0.4, NF vs. FC-0.6, NF vs. FC-0.8, FC-0.2 vs. FC-0.6

The detailed pairwise comparison results show consistent patterns across the three significant mechanical properties. For cube compressive strength (f_{cu}), the control group (NF) exhibited statistically significant differences when compared with fiber volume fractions of 0.4%, 0.6%, and 0.8%. Similarly, the FC-0.2 group showed significant differences compared to the FC-0.6 group. The mean differences ranged from 2.97 MPa to 5.08 MPa, all exceeding the HSD critical value of 2.9633 MPa.

For flexural strength (f_f), the same pattern emerged with the NF group showing significant differences relative to FC-0.4, FC-0.6, and FC-0.8 groups, and FC-0.2 versus FC-0.6 comparison also being significant. The mean differences ranged from 0.91 MPa to 1.37 MPa, surpassing the HSD critical value of 0.8329 MPa. Regarding axial compressive strength (f_{cp}), identical pairwise significant differences were observed. The mean differences varied from 2.79 MPa to 4.83 MPa, all exceeding the critical HSD value of 2.7593 MPa. The experimental results reveal that the FC-0.6 group (0.6% POM fiber volume fraction) achieved optimal performance across multiple mechanical properties. Compared to the control group (NF), significant improvements were observed as follows: cube compressive strength rose by 14.7% (34.49–39.57 MPa), flexural strength by 46.1% (2.97–4.34 MPa), and axial compressive strength by 15.5% (31.15–35.98 MPa).

The FC-0.4 and FC-0.8 groups also demonstrated significant improvements compared to the control group, indicating an effective fiber volume fraction range of 0.4% to 0.6%. However, the FC-1.0 group showed performance degradation compared to FC-0.6, suggesting that excessive fiber content may lead to diminished mechanical properties due to potential fiber agglomeration or matrix disruption.

The inferential statistical analysis confirms the reliability of the experimental findings. The low coefficient of variation values (<0.07) combined with the significant ANOVA results ($p < 0.01$ for three out of four properties) provide strong evidence for the effectiveness of POM fiber reinforcement in enhancing CAC mechanical properties. The splitting tensile strength, while showing numerical improvements, did not reach statistical significance ($p = 0.095$), indicating that POM fiber addition has limited impact on this particular mechanical property.

The comprehensive statistical analysis demonstrates that POM fiber incorporation significantly enhances the cube compressive strength, flexural strength, and axial compressive strength of CAC specimens, with the optimal fiber volume fraction identified as 0.6%. The Tukey HSD post hoc analysis provides clear evidence of significant performance differences between the control group and fiber-reinforced specimens, particularly at higher fiber volume fractions (0.4% to 0.8%). These findings provide a solid statistical foundation for optimizing POM fiber content in CAC formulations for enhanced mechanical performance.

5. Mechanical Test Results

5.1. Cube Compressive Property

Figure 4a reveals that as the POM fiber volume fraction increases, the average cube compressive strength of CAC specimens first rises and then declines, reaching a peak value of 39.6 MPa at a ρ value of 0.6, with a 95% confidence interval of 37.7 MPa to 41.5 MPa. Compared to specimens without POM fibers, the incorporation of POM fibers enhances the cube compressive strength of CAC, with a maximum improvement of 14.78%. This aligns with previous studies demonstrating that fiber reinforcement can increase concrete compressive strength. For instance, as illustrated in Figure 5, Liu et al. [35] observed that the compressive strength of CAC gradually improved with increasing carbon fiber content, exhibiting a typical columnar failure pattern during destruction, while the addition of carbon fibers reduced the number of large cracks in concrete. Dai et al. [36] tested the cube compressive strength of coral concrete under different contents of polypropylene fiber (PPF), glass fiber (GF), and basalt fiber (BF). The results showed that the addition of fiber enhanced the strength of CAC, mitigated its brittleness, and improved its ductility during failure. Notably, PPF exhibited the most significant improvement in concrete strength. This was attributed to the formation of a dense three-dimensional network structure by PPF during compression, which effectively inhibited crack propagation at appropriate dosage levels. Xu et al. and Zhang et al. [37,38] investigated the enhancement of compressive strength in recycled aggregate concrete (RAC) with basalt fibers and in ultra-high-performance concrete with polyvinyl alcohol fibers, respectively. Their studies further explain the reinforcement mechanisms of these fibers.

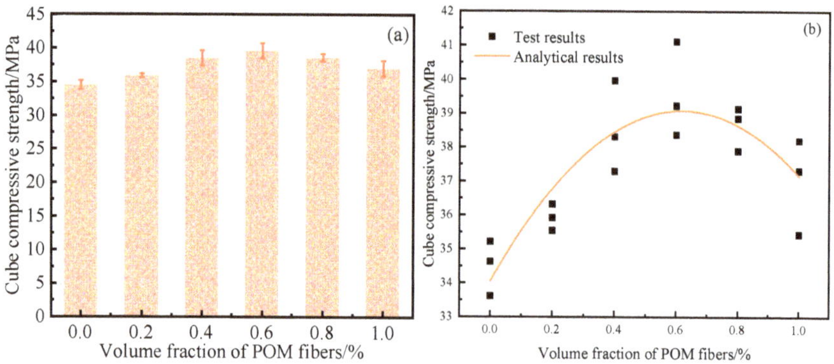

Figure 4. (**a**) Mean f_{cu} of CAC specimens with different ρ; (**b**) Comparison between analytical results and test results of f_{cu}.

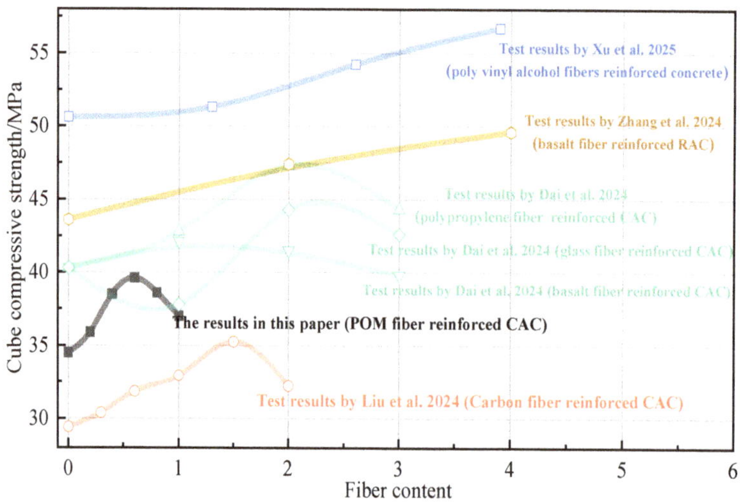

Figure 5. Comparison of compressive strength between POM-CAC and other types of fiber-reinforced concrete [35–38].

The polynomial fitting relationship between the compressive strength of coral concrete cubes and the fiber content ratio is shown in Equation (4). The comparison results between analytical results and test results of f_{cu} are shown in Figure 4b.

$$f_{cu} = -13.125\rho^2 + 16.225\rho + 34.050 \quad (4)$$

A comparative experimental study was conducted to investigate the effects of POM fibers on the compressive failure characteristics of CAC cubes. As shown in Figure 6a, the NF specimens exhibited complete fragmentation after compression testing, with extensive concrete spalling on the surface, demonstrating typical brittle failure behavior. Due to the low intrinsic strength of coral aggregates and strong aggregate–matrix interfacial bonding, all aggregates underwent penetrative fracture during failure, which significantly differs from the crack propagation pattern along aggregate interfaces observed in ordinary concrete. For specimens incorporating POM fibers (Figure 6b), surface spalling was still present but notably reduced, with only a few dominant cracks formed. As illustrated in Figure 6c, the fibers created effective bridging across vertical cracks. Compared to the control group, POM fibers enhanced material toughness by forming interfacial bonds that effectively connected adjacent cracks. Experimental results indicated that the addition of POM fibers improved the compressive integrity of specimens and delayed crack propagation through stress redistribution mechanisms.

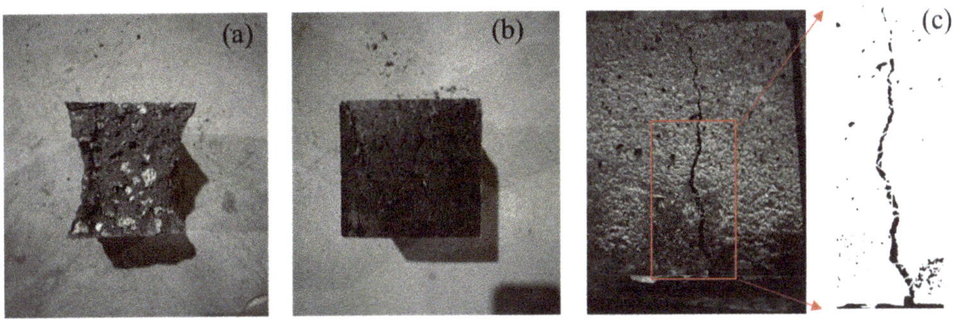

Figure 6. Failure performances of CAC specimens after the cube compressive test: (**a**) NF; (**b**) FC-0.6; (**c**) Bridging effect from POM fiber.

5.2. Splitting Tensile Property

As shown in Figure 7a, the average splitting tensile strength of CAC specimens first increases and then decreases with increasing POM fiber volume fraction. The maximum splitting tensile strength of CAC reaches 2.27 MPa at a POM fiber volume fraction of 0.6. Compared to specimens without POM fibers, CAC exhibits a nearly 20% improvement in splitting tensile strength, indicating significant mechanical enhancement. Polynomial fitting is applied to the experimental data, with the fitting results shown in Figure 7b.

Similar conclusions were reported by other researchers when strengthening the splitting compressive strength of concrete [16,37,39–41]. Figure 8 demonstrates that fibrillated polypropylene fibers effectively enhance both natural aggregate concrete (NAC) and RAC, improving their splitting tensile strength [39]. The addition of steel fibers during the reinforcement of recycled fine aggregate concrete (RFAC) or ordinary concrete can also moderately improve the splitting tensile performance of RFAC [40,41].

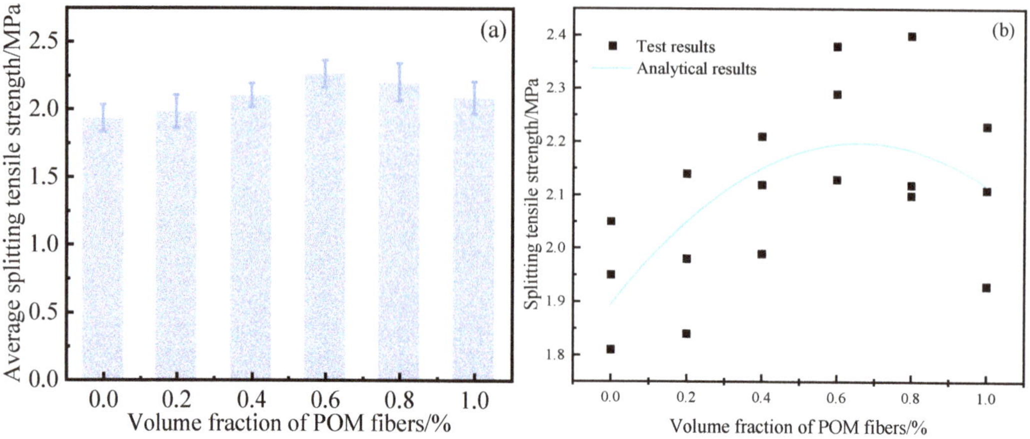

Figure 7. (**a**) f_{ts} of CAC specimens with different ρ; (**b**) Comparison between analytical and test results of f_{ts}.

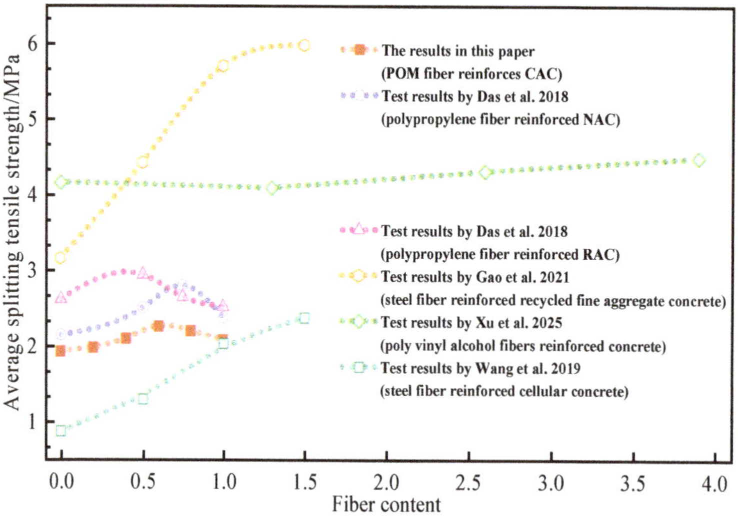

Figure 8. Comparison of splitting tensile between POM-CAC and other types of fiber-reinforced concrete [37,39–41].

The influence of POM fiber volume fractions on the splitting tensile failure morphology of CAC specimens is illustrated in Figure 9. As shown in Figure 9a, for CAC specimens without POM fibers, the specimen fractures first at the loading end section under sustained external load, with initial cracking and final failure occurring almost simultaneously, exhibiting typical brittle failure characteristics. Coral aggregates at the fracture surface show complete fracture patterns, where most particles form smooth fractures along the splitting plane, and some even display geometrically symmetric fracture patterns—a stark contrast to the fragmentation features of conventional concrete aggregates.

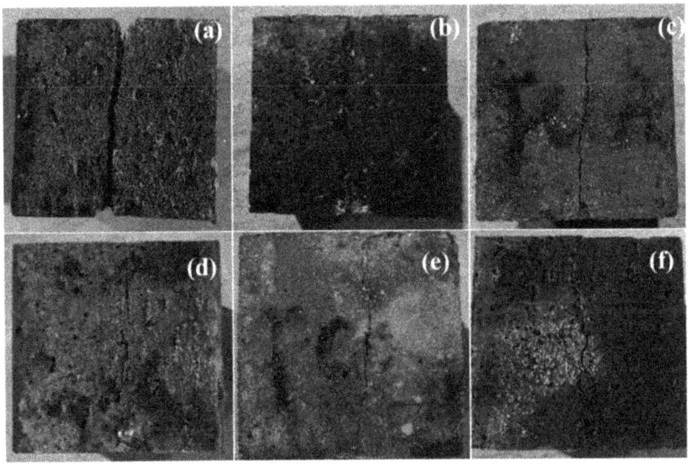

Figure 9. Failure performances of CAC specimens after the splitting tensile test: (**a**) NF; (**b**) FC-0.2; (**c**) FC-0.4; (**d**) FC-0.6; (**e**) FC-0.8; (**f**) FC-1.

In modified specimens containing POM fibers (Figure 9d), the material continues to bear partial load after initial crack formation before reaching peak stress, demonstrating a phased fracture process. At ultimate failure, crack width and propagation are significantly reduced, with fibers bridging cracks to form a mesh-like support structure. This maintains the specimen's geometric integrity at the macroscopic level, reflecting a typical stress redistribution mechanism. As shown in Figure 9f, when the fiber content increases to 1.0%, structural integrity is preserved, but crack width and quantity exceed those observed at 0.6% and 0.8% fiber volume fractions. This is attributed to excessive fiber content increasing material porosity and reducing specimen compactness. Experimental data indicate that a fiber volume fraction of 0.6–0.8% achieves optimal fiber–matrix synergy, enhancing material ductility while maintaining CAC structural density.

As stated in Principles of Reinforced Concrete [42], the empirical regression formula between the splitting tensile strength and cube compressive strength of standard ordinary concrete specimens is given by Equation (5):

$$f_{ts} = 0.19 f_{cu}^{3/4} \tag{5}$$

Previous studies revealed that Equation (3) does not apply rigorously to concrete incorporating POM fibers [43]. Therefore, based on prior research, the coefficient k in Equation (6) is modified as follows:

$$f_{ts} = k f_{cu}^{3/4} \tag{6}$$

The values of f_{cu}-mean, $f_{cu}^{3/4}$, and k are presented in Table 9. For varying POM fiber contents, k ranges from 0.1357 to 0.1438. Hence, k adopts the mean value of 0.1391 in Equation (6). The calculated splitting tensile strength values using Equation (6) are listed in Table 9. The ratio of tested to calculated splitting tensile strengths fluctuates between 0.98 and 1.03, demonstrating that the modified empirical formula for the relationship between splitting tensile strength and cube compressive strength provides reliable reference.

Table 9. Ratio of test splitting tensile strength to calculated splitting strength.

ρ	0.00	0.20	0.40	0.60	0.80	1.00
f_{cu}-mean	34.50	35.90	38.50	39.60	38.60	37.00
$f_{cu}^{3/4}$	14.24	14.67	15.46	15.79	15.49	15.00
k	0.1363	0.1357	0.1365	0.1438	0.1427	0.1393
Calculated f_t	1.98	2.04	2.15	2.20	2.15	2.09
Tested f_{ts}/Calculated f_{ts}	0.98	0.98	0.98	1.03	1.03	1.00

5.3. Flexural Property

As shown in Figure 10, the average flexural strength of CAC specimens first increases and then decreases with increasing POM fiber volume fraction. The maximum flexural strength of CAC reaches 4.34 MPa at a POM fiber volume fraction of 0.6, with a 95% confidence interval of 3.87–4.80 MPa. Compared to specimens without POM fibers, CAC exhibits a significant improvement in flexural strength. Polynomial fitting was applied to the experimental data, with the fitting results illustrated in Figure 10.

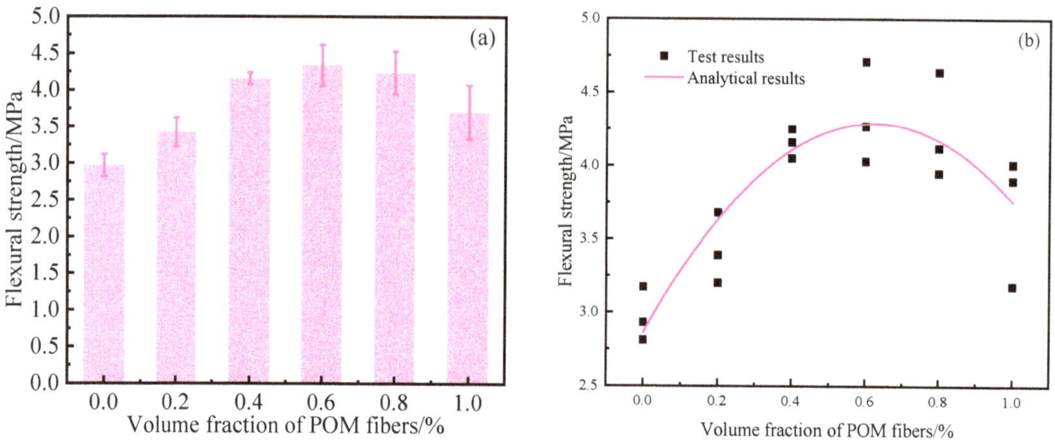

Figure 10. (**a**) f_f of CAC specimens with different ρ; (**b**) Comparison between analytical and test results of f_f.

Figure 11 illustrates the influence of POM fibers on the flexural failure morphology of CAC specimens. For NF specimens, no visible cracks appear before peak load, but sudden fracture occurs at ultimate failure, with cracks instantly penetrating the cross-section. Coral aggregates at the fracture surface display geometrically symmetric fracture patterns, distinct from the random fragmentation observed in conventional concrete aggregates. In specimens containing POM fibers, residual load-bearing capacity maintains 15–46% of the peak load during the initial flexural stage due to the interfacial bonding effect between POM fibers and the matrix, demonstrating pronounced ductile fracture characteristics. At final failure, the main crack width is reduced owing to the bonding effect of POM fibers. Based on flexural test results and post-fracture morphological analysis, specimens with a POM fiber volume fraction of 0.6% exhibit optimal crack resistance performance, showing significantly slower crack propagation rates compared to the NF group and no secondary crack formation. When the fiber content increases to 0.8%, structural integrity is retained, but crack width and length increase, indicating that excessive fiber content causes localized fiber clustering, compromising matrix continuity. Thus, an appropriate POM fiber volume fraction effectively delays the development of initial microcracks and inhibits new crack formation, achieving excellent crack resistance.

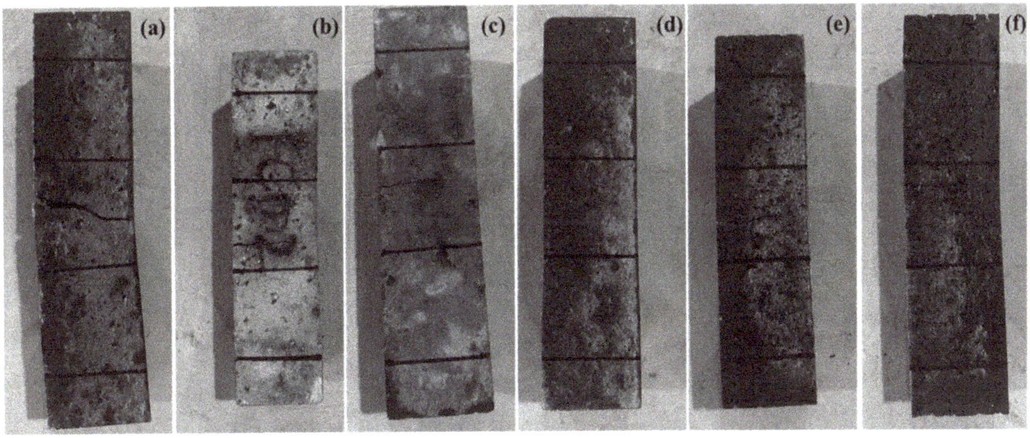

Figure 11. Failure performances of CAC specimens after the flexural test: (**a**) NF; (**b**) FC-0.2; (**c**) FC-0.4; (**d**) FC-0.6; (**e**) FC-0.8; (**f**) FC-1.

Based on previous studies [43,44], the relationship between flexural strength and cube compressive strength is hypothesized as shown in Equation (7). The fitting results are presented in Figure 12. In Equation (7), the values of a and b are 1.6486×10^{-4} and 2.7735, respectively.

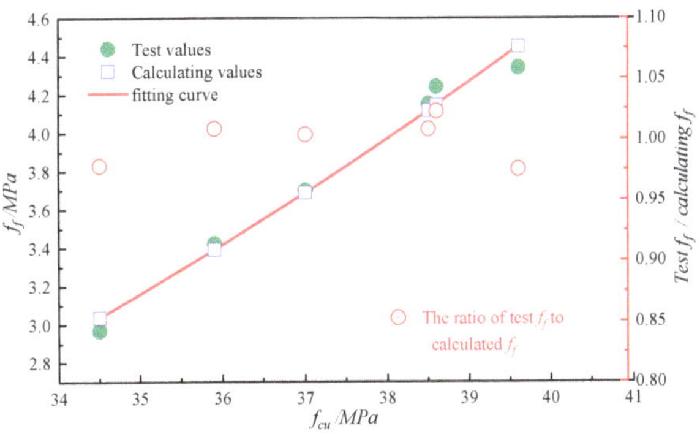

Figure 12. Test flexural strength, calculate flexural strength, and the ratio between them.

Using Equation (7), the flexural strength of CAC specimens at different POM fiber volume fraction was calculated. The ratio of experimentally tested flexural strength to calculated values is shown in Table 10 and Figure 12. The ratio fluctuates between 0.98 and 1.01, indicating that the modified empirical formula can effectively predict the flexural strength of coral aggregate concrete based on cube compressive strength. Therefore, the empirical formula relating flexural strength to cube compressive strength for polyoxymethylene fiber-reinforced coral concrete is given by Equation (8).

$$f_f = a f_{cu}^{b} \tag{7}$$

$$f_f = 0.000016486 f_{cu}^{2.7735} \tag{8}$$

Table 10. Test flexural strength, calculating flexural strength, and the ratio between them.

ρ	0.00	0.20	0.40	0.60	0.80	1.00
f_{cu}-mean	34.50	35.90	38.50	39.60	38.60	37.00
f_f	2.97	3.42	4.15	4.34	4.24	3.70
Calculating f_f	3.04	3.39	4.12	4.45	4.14	3.69
Tested f_f/Calculated f_f	0.98	1.01	1.01	0.98	1.02	1.00

5.4. Axial Compressive Property

The average axial compressive strength of CAC specimens is shown in Figure 13a. A polynomial regression analysis was subsequently conducted on the experimental results, with the fitting results illustrated in Figure 13b.

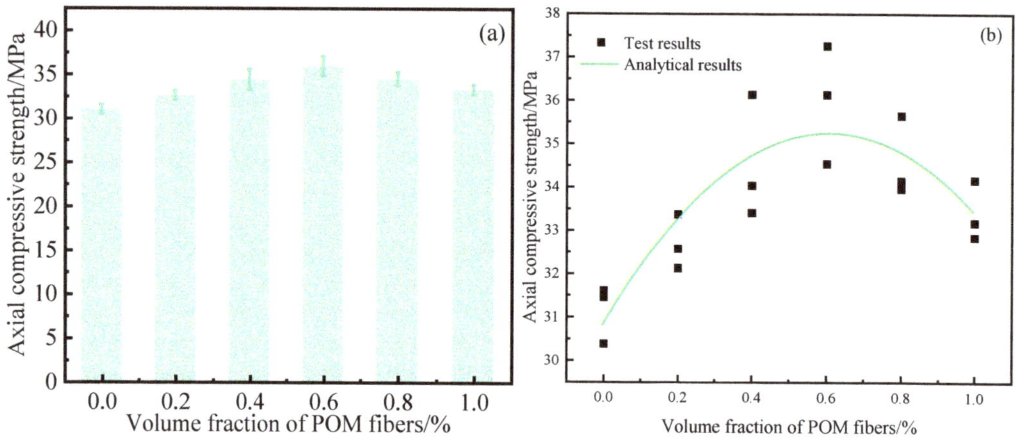

Figure 13. (**a**) f_{cp} of CAC specimens with different ρ; (**b**) Comparison between analytical and test results of f_{cp}.

As illustrated in Figure 13, the average axial compressive strength of CAC specimens initially increases and then decreases with increasing POM fiber volume fraction, mirroring the trend observed for cube compressive strength. At a POM fiber volume fraction of 0.6%, the peak axial compressive strength reaches 35.98 MPa, representing a 4.83 MPa improvement compared to NF.

Figure 14 demonstrates the distinct failure morphologies of CAC specimens under axial compression across varying fiber volume fractions. The NF group exhibited vertical microcracks during initial loading, which progressively developed into diagonal cracks, culminating in extensive concrete spalling and through-aggregate fractures—a hallmark of brittle failure. In contrast, POM fiber-reinforced specimens maintained post-failure integrity through fiber-mediated interlocking, with significantly reduced spalling volume and fragment count compared to NF. Partial fragments remained anchored by fibers, showcasing a characteristic fiber–matrix interfacial bond failure. The failure mechanisms of axial compressive specimens align with those of cubic specimens: POM fibers bridging vertical cracks effectively delayed microcrack propagation and suppressed macrocrack formation. Optimal structural integrity was achieved at 0.6% and 0.8% fiber volume fractions, with the FC0.8 group exhibiting diffuse crack patterns—a stark contrast to NF's localized failure—attributed to fiber-induced mitigation of stress concentration. However, fiber dosages exceeding 0.8% reduced matrix densification, reintroducing diagonal cracking patterns akin to NF. These findings confirm that moderate POM fiber incorporation (0.6–0.8%)

significantly enhances CAC's ductility, while excessive dosages compromise performance. Through this toughening mechanism, catastrophic through-aggregate fractures in coral concrete are suppressed, enhancing synergistic load-bearing capacity between aggregates and the matrix.

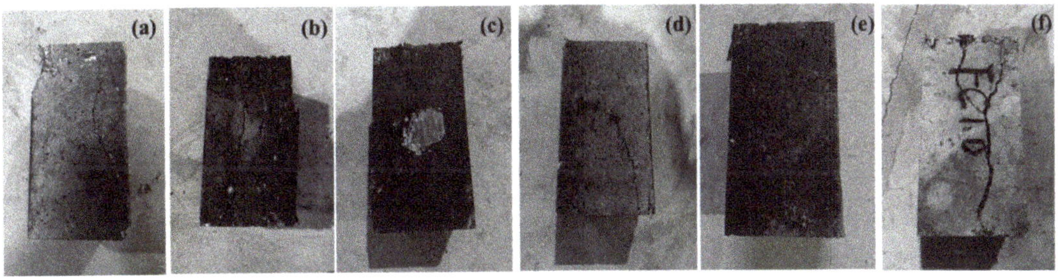

Figure 14. Failure performances of CAC specimens after the axial compressive test: (**a**) NF; (**b**) FC-0.2; (**c**) FC-0.4; (**d**) FC-0.6; (**e**) FC-0.8; (**f**) FC-1.

5.5. The CAC Constitutive Model

The stress–strain full curve, the maximum strain ε_c and maximum stress σ_c at failure, and the dimensionless stress–strain curve of the CAC blocks under different POM fiber content are shown in Figure 15. After the sample failure, the splitting test was stopped, so the stress–strain curve only shows the ascending part of the curve.

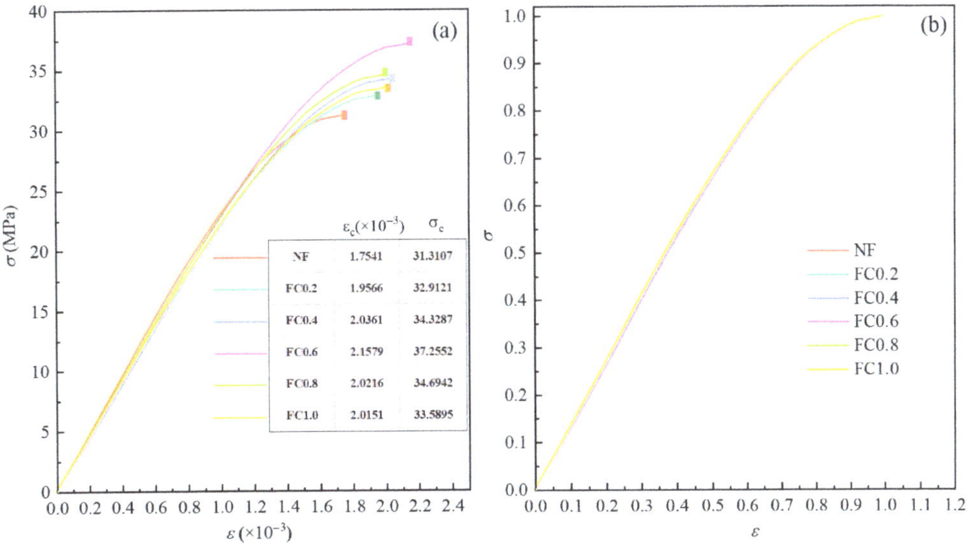

Figure 15. Uniaxial compression stress–strain curves of CAC specimens: (**a**) Complete stress–strain curves with corresponding ε_c and σ_c at failure; (**b**) Normalized stress–strain curves.

In the early stage of pressure loading, no initial microcracks appeared on the concrete surface, and the stress–strain curves of different blocks showed no significant difference, remaining in the linear elastic deformation stage. As the pressure increased, many microcracks were generated on the surface of the block, extending from the loading end downward and internally. The surface of the block showed no obvious cracks, and the slope of the curve gradually decreased. When the stress reached 0.85 σ_c [45], the specimen exhibited noticeable lateral deformation and surface cracks, with crack development showing anisotropy. Due to the bridging effect of the POM fibers, compared with the NC group CAC specimen, the load-bearing capacity of the CAC specimen increased, the width of the surface cracks was smaller, and the curve remained relatively flat for a longer time before reaching the maximum failure stress.

Coral aggregates exhibit inherently low strength and pronounced brittle deformation characteristics. The incorporation of POM fibers enhances both the strength and toughness of the concrete. Due to its distinct deformation behavior compared to ordinary concrete, it is necessary to develop a constitutive model applicable to CAC containing POM fibers.

Based on existing research [43], the constitutive model of the coral aggregate concrete specimens was finally selected using the Guozhenhai model, the Sargin model, and the Carreira and Chu model to fit the stress–strain curves. The expressions for these curve equations are given in Equations (9)–(11), respectively.

$$\text{Guozhenhai model} \quad y = ax + (3-2a)x^2 + (a-2)x^3 \tag{9}$$

$$\text{Sargin model} \quad y = \frac{c_1x+(c_2-1)x^2}{1+c_1x-2x+c_2x^2} \tag{10}$$

$$\text{Carreira and Chu model} \quad y = \frac{nx}{n-1+x^n} \tag{11}$$

In the equations, $x = \varepsilon/\varepsilon_c, y = \varepsilon/\varepsilon_c$, where a, c_1, c_2, and n are constants specific to each model.

The nonlinear fitting of the dimensionless stress–strain curves of CAC specimens was performed using the above three models, yielding the governing constants a, c_1, c_2, and n for each constitutive equation. The computational results are presented in Table 11 and Figure 16. As shown in Figure 16, the value of a in the Guozhenhai model decreases initially then increases with increasing ρ; in the Sargin model, c_1 first increases and then decreases with the value of ρ, while c_2 first decreases and then increases. The variation in the control constant n in the Carreira and Chu model follows the same trend as parameter a in the Guozhenhai model. Table 11 demonstrates that all three models achieve the value of R^2 exceeding 0.99 with experimental data, indicating excellent agreement and enabling accurate prediction of the stress–strain behavior in CAC specimens. R^2 represents the goodness of fit of a regression model, calculated using Equation (12):

$$R^2 = 1 - \frac{\sum_{i=1}^{n}(y_i - \hat{y}_i)^2}{\sum_{i=1}^{n}(y_i - \overline{y})^2} \tag{12}$$

where y_i is the i-th observed value, $\overline{y}$ is the mean of the observed values, $\hat{y}_i$ is the i-th predicted value, and n is the number of data points.

Table 11. Parameters and fitting performance of different constitutive models.

Model	Parameter	NC	FC0.2	FC0.4	FC0.6	FC0.8	FC1.0
Guozhenhai	a	1.258	1.236	1.222	1.250	1.277	1.295
	R^2	0.99998	0.99998	0.99998	0.99998	0.99998	0.99994
Sarging	c_1	1.291	1.268	1.254	1.279	1.307	1.319
	c_2	0.327	0.355	0.347	0.341	0.302	0.328
	R^2	0.99996	0.99995	0.99995	0.99996	0.99996	0.99994
Carreira and Chu	n	3.900	4.091	4.120	3.975	3.875	3.821
	R^2	0.99988	0.99991	0.99989	0.99991	0.99991	0.99990

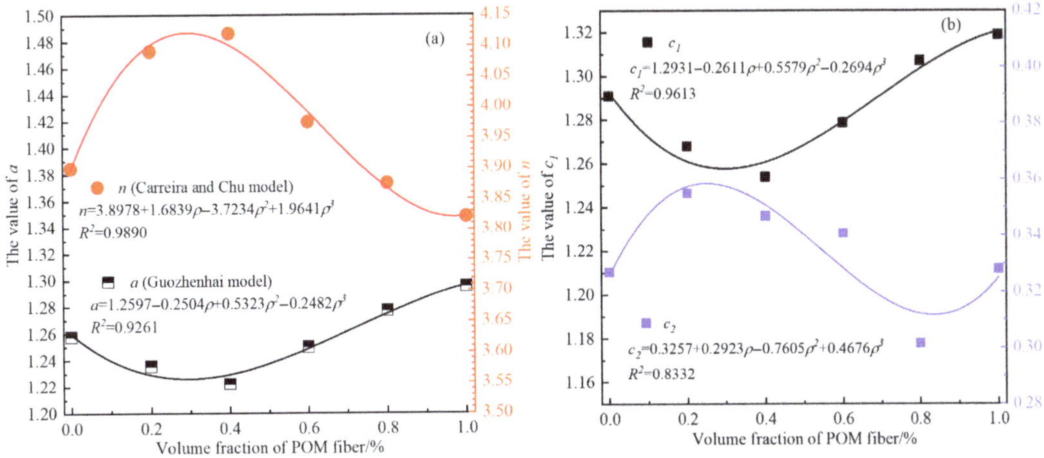

Figure 16. Functional relationships between model parameters and ρ with corresponding R^2 values: (**a**) Guozhenhai model and Carreira and Chu model; (**b**) Sargin model.

Due to the strong correlation between the governing constants of the model parameters and ρ, functional relationships were established for constants a, c_1, c_2, and n in the three constitutive equations. As shown in Figure 16b, the Sargin model involves two parameters with R^2 values of 0.9613 and 0.8332 against ρ. Consequently, the Sargin model was excluded primarily due to its multi-parameter dependency. Figure 16a demonstrates that both the Guozhenhai and Carreira and Chu models employ a single governing parameter (a and n, respectively). The R^2 values between ρ and the governing parameters a (Guozhenhai model) and n (Carreira and Chu model) are 0.9261 and 0.9890, respectively. Owing to its significantly superior correlation, the Carreira and Chu model was selected as the optimal constitutive framework for POM fiber-reinforced CAC. The functional relationship between n and ρ is given by Equation (13). Substituting Equation (13) into Equation (13) yields the modified Carreira and Chu model, expressed as Equation (14).

$$a = 3.8978 + 1.6839\rho - 3.7234\rho^2 + 1.9641\rho^3 \tag{13}$$

$$y = \frac{\left(3.8978 + 1.6839\rho - 3.7234\rho^2 + 1.9641\rho^3\right)x}{2.8978 + 1.6839\rho - 3.7234\rho^2 + 1.9641\rho^3 + x^{(3.8978+1.6839\rho-3.7234\rho^2+1.9641\rho^3)}} \tag{14}$$

5.6. The Model Validation

The experimental stress–strain data of CAC specimens under varying POM fiber contents and the corresponding predictions of the modified Carreira and Chu model are presented in Figure 17, with residuals shown in Figure 18. The fitted curves for all test groups achieve R^2 values exceeding 0.9999, indicating near-perfect agreement between model predictions and experimental results. Calculated curves align closely with measured data points without discernible deviations, demonstrating consistent accuracy across different fiber volume fractions.

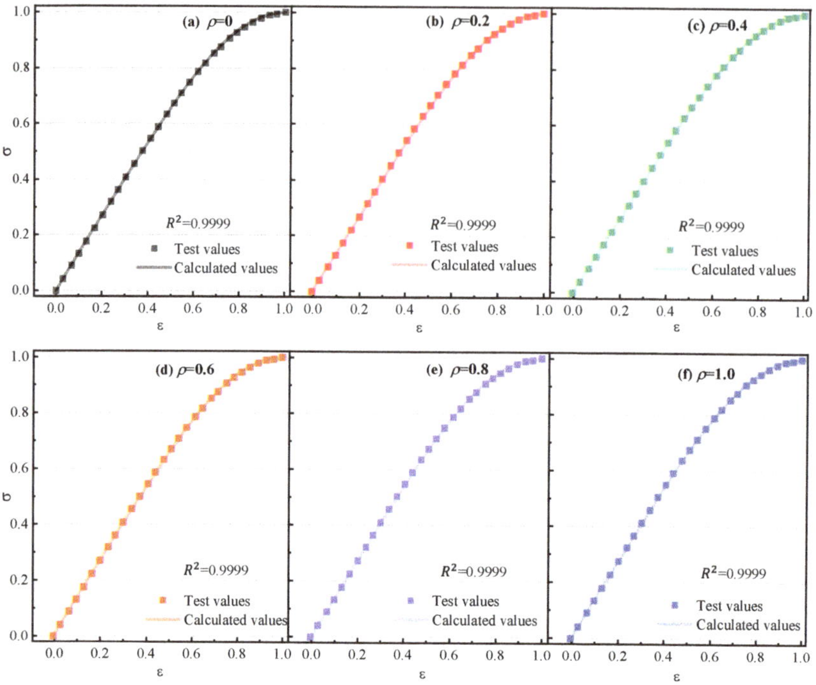

Figure 17. Comparisons of experimental results with the calculated values originated from the constitutive model.

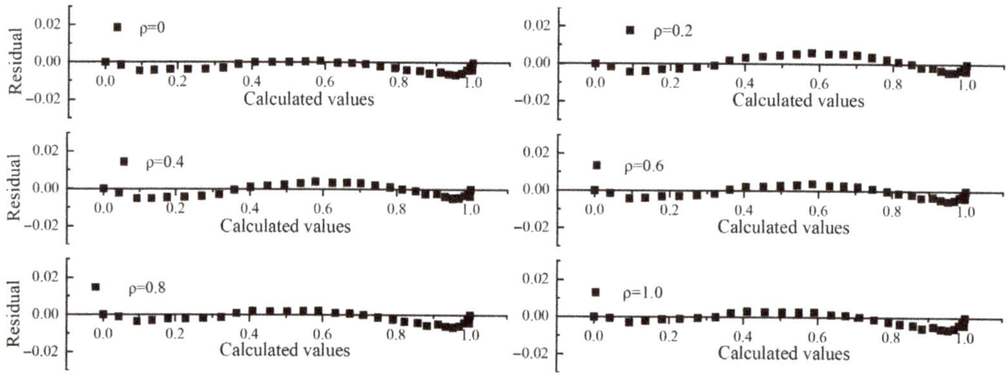

Figure 18. The residual plots of the modified Carreira and Chu model.

In Figure 18, residuals are randomly distributed about the zero-reference line with no systematic trends. The maximum absolute residual is approximately 0.02—a negligible deviation relative to the stress magnitude. Critically, residual distribution patterns remain consistent across all POM fiber contents, confirming stable error control despite parameter variations.

In summary, the model exhibits exceptional performance in both statistical goodness-of-fit and engineering error stability, providing a highly accurate representation of the stress–strain behavior of POM fiber-reinforced CAC specimens.

6. Conclusions

This study investigates the effect of fiber content on the early mechanical properties of coral aggregate concrete (CAC) by preparing specimens containing varying volume fractions of POM fibers. It proposes the optimal fiber content range and establishes a constitutive model suitable for CAC specimens. The main conclusions are as follows:

(1) POM fibers reduce the slump and slump flow of CAC. With increasing POM fiber volume fraction, the cube compressive strength, axial compressive strength, split tensile strength, and flexural strength of CAC first increase and then decrease. All strength values peak at a POM fiber volume fraction of 0.6%, reaching 39.6 MPa, 35.98 MPa, 2.27 MPa, and 4.34 MPa, respectively. Compared to the NF group, these properties improved by 14.78%, 15.50%, 17.01%, 46.13%, and 3.69%, respectively. POM fiber incorporation significantly enhances the early mechanical properties of CAC specimens.

(2) At the optimal ρ, POM fibers distribute uniformly within the specimens, forming a three-dimensional network structure that suppresses crack propagation and redistributes stress. The interfacial bonding between fibers and the matrix delays microcrack initiation, reduces the width of primary cracks, and transforms the failure mode of CAC from brittle to ductile. Additionally, fiber addition lowers specimen porosity and strengthens the synergistic load-bearing capacity between aggregates and the cementitious matrix. When the POM fiber volume fraction exceeds 0.8%, fiber agglomeration increases localized porosity and creates interfacial weak zones, which act as stress concentration points and degrade mechanical performance. Thus, the optimal synergistic effect is achieved within the fiber dosage range of 0.6–0.8%.

(3) Based on the stress–strain curves under different POM fiber contents, a modified Carreira–Chu constitutive model incorporating volume fraction (ρ) was established. The goodness-of-fit (R^2) for the entire stress process of CAC exceeded 0.99, enabling accurate prediction of its mechanical behavior. This provides a reliable theoretical tool for engineering design and technical support for the efficient conversion and green application of coral aggregates—a low-carbon resource—in marine engineering.

This study only analyzed the influence of POM fibers on the mechanical properties of coral concrete at 28 days of age. Further research can be conducted to study the durability, bonding performance, and shrinkage performance of CAC with POM fiber in marine environments, providing more comprehensive data support for the engineering application of CAC. Furthermore, within the framework of relevant administrative regulations, further investigation is needed to explore the possibility of achieving sustainable development of marine resources and environment by utilizing local materials and properly managing coral debris waste.

Author Contributions: J.W.: Writing—original draft, Conceptualization, Investigation, Methodology, Data curation, Supervision, Funding acquisition; W.S.: Writing—original draft, Writing—review and editing, Formal analysis, Supervision. L.T.: Writing—original draft, Writing—review and editing, Formal analysis, Supervision. All authors have read and agreed to the published version of the manuscript.

Funding: This article was Supported by the Key project of science and technology research program of Chongqing Education Commission of China (KJZD-K202503205).

Data Availability Statement: The original contributions presented in the study are included in the article. Further inquiries can be directed to the corresponding author.

Conflicts of Interest: The authors declare no conflicts of interest.

References

1. Cui, Y.; Jiang, J.; Fu, T.; Liu, S. Feasibility of using Waste Brine/Seawater and Sea Sand for the Production of Concrete: An Experimental Investigation from Mechanical Properties and Durability Perspectives. *Sustainability* **2022**, *14*, 13340. [CrossRef]
2. Ganesan, K.; Kanagarajan, V.; Dominic, J.R.J. Influence of marine sand as fine aggregate on mechanical and durability properties of cement mortar and concrete. *Mater. Res. Express* **2022**, *9*, 035504. [CrossRef]
3. Hayek, M.; Soleimani, T.; Salgues, M.; Souche, J.C. Utilizing Fine Marine Sediment as a Partial Substitute for Sand in Self-Compacting Concrete Specially Designed for Application in Marine Environments. *Sustainability* **2024**, *16*, 2538. [CrossRef]
4. Wang, A.; Lyu, B.; Zhang, Z.; Liu, K.; Xu, H.; Sun, D. The development of coral concretes and their upgrading technologies: A critical review. *Constr. Build. Mater.* **2018**, *187*, 1004–1019. [CrossRef]
5. Sun, L.; Wang, C.; Zhang, C.; Yang, Z.; Li, C.; Qiao, P. Experimental investigation on the bond performance of sea sand coral concrete with FRP bar reinforcement for marine environments. *Adv. Struct. Eng.* **2022**, *26*, 533–546. [CrossRef]
6. Sun, L.; Yang, Z.; Qin, R.; Wang, C. Mix design optimization of seawater sea sand coral aggregate concrete. *Sci. China Technol. Sci.* **2022**, *66*, 378–389. [CrossRef]
7. Zhang, B.; Zhu, H.; Dong, Z.; Yang, Z. Mechanical properties and durability of FRP-reinforced coral aggregate concrete structures: A critical review. *Mater. Today Commun.* **2023**, *35*, 105656. [CrossRef]
8. Zhang, X.; Niu, D. Dynamic compressive behavior of hybrid basalt-polypropylene fibers reinforced coral aggregate concrete. *J. Build. Eng.* **2024**, *95*, 110291. [CrossRef]
9. Wang, Y.; Shui, Z.; Gao, X.; Huang, Y.; Yu, R.; Li, X. Utilizing coral waste and metakaolin to produce eco-friendly marine mortar: Hydration, mechanical properties and durability. *J. Clean. Prod.* **2019**, *219*, 763–774. [CrossRef]
10. Chen, Y.; Li, P.; Zhang, S. Experimental investigation on triaxial mechanical properties of coral coarse aggregate-sea sand seawater concrete. *Constr. Build. Mater.* **2023**, *409*, 134213. [CrossRef]
11. Da, B.; Huang, H.; Tao, T.; Chen, Y.; Chen, D. Experimental and numerical investigations on the mechanical properties of coral aggregate seawater concrete. *Eng. Fract. Mech.* **2024**, *310*, 110498. [CrossRef]
12. Guo, R.; Dong, J.; Ma, L.; Long, Z.; Xu, F.; Yin, C. Three-dimensional mesoscopic investigation on the dynamic compressive behavior of coral sand concrete with reinforced granite coarse aggregate (GCA). *Compos. Struct.* **2025**, *352*, 118650. [CrossRef]
13. Huang, Y.; Cheng, Z.; Dong, L.; Cao, P.; Zhong, J.; Wang, Q.; Loan, P. Experimental and analytical studies on the improvements in mechanical properties of coral aggregates concrete through aggregates strengthening method. *Constr. Build. Mater.* **2024**, *438*, 137243. [CrossRef]
14. Cao, S.; Zheng, D.; Yilmaz, E.; Yin, Z.; Xue, G.; Yang, F. Strength development and microstructure characteristics of artificial concrete pillar considering fiber type and content effects. *Constr. Build. Mater.* **2020**, *256*, 119408. [CrossRef]
15. Shafiei, S.; Mohammadi, Y. Effect of AR-glass fibers and natural micronized zeolite on mechanical and permeability properties of concrete exposed to normal, sodium chloride, and magnesium sulfate environments. *Constr. Build. Mater.* **2024**, *457*, 139288. [CrossRef]
16. Zhu, L.; Wen, T.; Tian, L. Size effects in compressive and splitting tensile strengths of polypropylene fiber recycled aggregate concrete. *Constr. Build. Mater.* **2022**, *341*, 127878. [CrossRef]
17. Shi, F.; Yin, S.; Pham, T.M.; Tuladhar, R.; Hao, H. Pullout and flexural performance of silane groups and hydrophilic groups grafted polypropylene fibre reinforced UHPC. *Constr. Build. Mater.* **2021**, *277*, 122335. [CrossRef]
18. Ye, J.; Cui, C.; Yu, J.; Yu, K.; Dong, F. Effect of polyethylene fiber content on workability and mechanical-anisotropic properties of 3D printed ultra-high ductile concrete. *Constr. Build. Mater.* **2021**, *281*, 122586. [CrossRef]
19. Adebayo, I.W.; Long, G.; Tang, Z.; Ghone, M.O.; Zaland, S.; Garba, M.J.; Yang, K.; Akhunzada, K.; Oluwasina, U.A. Effect of crumb rubber and polyethylene fiber on the strength and toughness of fly ash/slag-based geopolymer concrete. *Constr. Build. Mater.* **2024**, *455*, 139133. [CrossRef]
20. Guo, W.; Ding, Y.; Randl, N. Effect of macro polyoxymethylene fibers on shear transfer across a crack and crack surface topography of concrete. *Constr. Build. Mater.* **2025**, *464*, 140118. [CrossRef]
21. Zhang, J.; Li, B.; Chen, W.; Guo, R. Experimental investigations on tensile strength and fracture toughness of a polyoxymethylene fiber reinforced concrete. *Theor. Appl. Fract. Mec.* **2024**, *130*, 104250. [CrossRef]
22. Hua, J.; Yang, Z.; Xue, X.; Huang, L.; Wang, N.; Chen, Z. Bond properties of bimetallic steel bar in seawater sea-sand concrete at different ages. *Constr. Build. Mater.* **2022**, *323*, 126539. [CrossRef]
23. Wang, F.; Xue, X.; Hua, J.; Wang, N.; Yao, Y. Properties of polyoxymethylene fibre-reinforced seawater sea sand concrete exposed to high temperatures. *Constr. Build. Mater.* **2023**, *409*, 133854. [CrossRef]
24. Yan, Y.; Guo, R.; Wei, Y.; Pan, T.; Li, X.; Tan, L. Strain-rate sensitivity analysis of the fracture behaviour of polyoxymethylene-fibre-reinforced airport-pavement concrete. *Constr. Build. Mater.* **2023**, *402*, 133048. [CrossRef]
25. Xue, X.; Wang, F.; Hua, J.; Wang, N.; Huang, L.; Chen, Z.; Yao, Y. Effects of Polyoxymethylene Fiber on Fresh and Hardened Properties of Seawater Sea-Sand Concrete. *Polymers* **2022**, *14*, 4969. [CrossRef]

26. Wang, F.; Hua, J.; Xue, X.; Wang, N.; Yao, Y. Effects of Polyoxymethylene Fiber on Mechanical Properties of Seawater Sea-Sand Concrete with Different Ages. *Polymers* **2022**, *14*, 3472. [CrossRef]
27. Dong, J.; Zong, Y.; Shang, X.; Chen, X.; Tu, Z.; Jiang, R.; Zhu, Z. Impact of Various Erosive Environments on the Durability of POM Fiber-Reinforced Ultra-High-Performance Concrete. *Buildings* **2024**, *14*, 4048. [CrossRef]
28. Ding, Y.; Guo, W.; Li, D.; Pacheco-Torgal, F. Exploring the effect of polyoxymethylene fiber on concrete toughness and self-sensing capability of concrete cracking under bending. *Constr. Build. Mater.* **2024**, *454*, 138933. [CrossRef]
29. *GB/T 50081-2019;* Standard for Test Method of Mechanical Properties on Ordinary Concrete. Ministry of Housing and Urban-Rural Development of the People's Republic of China and State Administration for Market Regulation: Beijing, China, 2019.
30. *GB/T 17431.2-2010;* Lightweight Aggregates and Its Test Methods—Part 2: Test Methods for Lightweight Aggregates. General Administration of Quality Supervision, Inspection and Quarantine of the People's Republic of China and Standardization Administration of the People's Republic of China: Beijing, China, 2010.
31. *GB/T 21120-2018;* Synthetic Fibres for Cement Concrete and Mortar. State Administration for Market Regulation and Standardization Administration of the People's Republic of China: Beijing, China, 2018.
32. *GB 8076-2008;* Concrete Admixtures. General Administration of Quality Supervision, Inspection and Quarantine of the People's Republic of China and Standardization Administration of the People's Republic of China: Beijing, China, 2008.
33. *JGJ/T 12-2019;* Technical Standard for Application of Lightweight Aggregate Concrete. Ministry of Housing and Urban-Rural Development of the People's Republic of China: Beijing, China, 2019.
34. *GB/T 50080-2016;* Standard for Test Method of Performance on Ordinary Fresh Concrete. Ministry of Housing and Urban-Rural Development of the People's Republic of China and General Administration of Quality Supervision, Inspection and Quarantine of the People's Republic of China: Beijing, China, 2016.
35. Liu, B.; Wang, F.; Zou, C.; Ming, Y.; Qin, J.; Qian, K. Comparison of the mechanical properties and constitutive models of carbon fiber-reinforced coral concrete cubes and prisms under uniaxial compression. *J. Build. Eng.* **2024**, *98*, 111247. [CrossRef]
36. Dai, J.; Yin, S.; Wang, F.; Ge, M. Analysis of microstructure and compressive properties of full coral concrete under fiber reinforcement. *J. Build. Eng.* **2024**, *93*, 109826. [CrossRef]
37. Xu, Q.; Jiang, X.; Zhang, Z.; Xu, C.; Zhang, J.; Zhou, B.; Hang, W.; Zheng, Z. Experimental study on residual mechanical properties of steel-PVA hybrid fiber high performance concrete after high temperature. *Constr. Build. Mater.* **2025**, *458*, 139735. [CrossRef]
38. Zhang, X.; Zhu, Y.; Shen, Y.; Wang, J.; Fan, Y.; Gao, X.; Huang, Y. Compressive mechanical performance and microscopic mechanism of basalt fiber-reinforced recycled aggregate concrete after elevated temperature exposure. *J. Build. Eng.* **2024**, *96*, 110647. [CrossRef]
39. Das, C.S.; Dey, T.; Dandapat, R.; Mukharjee, B.B.; Kumar, J. Performance evaluation of polypropylene fibre reinforced recycled aggregate concrete. *Constr. Build. Mater.* **2018**, *189*, 649–659. [CrossRef]
40. Gao, D.; Wang, F. Effects of recycled fine aggregate and steel fiber on compressive and splitting tensile properties of concrete. *J. Build. Eng.* **2021**, *44*, 102631. [CrossRef]
41. Wang, X.; Zhang, S.; Wang, C.; Cao, K.; Wei, P.; Wang, J. Effect of steel fibers on the compressive and splitting-tensile behaviors of cellular concrete with millimeter-size pores. *Constr. Build. Mater.* **2019**, *221*, 60–73. [CrossRef]
42. Guo, Z.; Shi, X. *Principles of Reinforced Concrete;* Tsinghua University Press: Beijing, China, 2013.
43. Yao, Y. *Study on Mechanical Properties and High Temperature Resistance of Polyoxymethylene Fiber Seawater and Sea-Sand Concrete;* Chongqing University: Chongqing, China, 2022.
44. Cai, Z.; Li, S. The Experimental Correlation Between Flexural Tensile Strength and Flexural Compressive Strength of Cement Concrete Pavement. *China J. Highw. Transp.* **1992**, *5*, 14–20.
45. Cui, T.; He, H.; Yan, W. Experimental Research on Uniaxial Compression Constitutive Model of Hybrid Fiber-Reinforced Cementitious Composites. *Materials* **2019**, *12*, 2370. [CrossRef]

buildings

MDPI

Article

Strain Energy-Based Calculation of Cracking Loads in Reinforced Concrete Members

Tao Zheng and Gui-Yao Wang *

School of Civil and Environmental Engineering, Changsha University of Science and Technology, Changsha 410114, China; zhengtao@stu.csust.edu.cn
* Correspondence: 19002010035@csust.edu.cn

Abstract

The traditional methods for calculating the cracking loads of concrete members require the introduction of a semi-empirical inelastic influence coefficient of the section resistance moment to reflect the influence of sectional inelastic deformation development, and its value needs to be corrected according to the sectional characteristics and material properties. In contrast, emerging machine learning models for predicting the cracking loads of concrete members lack clear mechanical mechanisms, making their predictions difficult to interpret. Based on this, member cracking is reinterpreted as a change in the macroscopic performance of the member, and the strain energy variation before and after the fracture of plain concrete axially tensioned members is analyzed. A viewpoint is proposed that employs the incremental change in strain energy to characterize member cracking. With the internal tensile deformation representing the total strain energy of the member, a calculation method for the member cracking loads is established based on the critical condition that the second-order differential of this strain energy with respect to the member deformation equals zero. This method is applicable to both axially tensioned and flexural members of steel- and FRP-reinforced concrete that primarily undergo axial tensile deformation. The method is applied to analyze the cracking of axially tensioned and flexural members under different concrete strength grades, reinforcement types, and reinforcement ratios. It successfully explains the mechanism whereby the strain in the tension zone at member cracking exceeds the material ultimate tensile strain. Calculations verified against experimental data from four sets of tests on steel- or FRP-reinforced concrete beams demonstrate that the proposed method can be accurately applied to both normal- and high-strength concrete, as well as to both steel and FRP reinforcement, with a relative error of only 1% and a coefficient of variation of 0.12.

Keywords: fracture; cracking loads; reinforced concrete member; strain energy

1. Introduction

In recent years, with the widespread application of high- and ultra-high-strength concrete, as well as new reinforcement materials such as fiber-reinforced polymers (FRPs) and eco-friendly recycled materials [1–9], the sectional stress and strain in steel- and FRP-reinforced concrete members under serviceability limit states have increased significantly. In some cases, the deformation and crack development of such concrete members may even exceed the limits specified in design codes, thereby affecting the normal serviceability of the structure. Consequently, the crack resistance of members at the serviceability stage has increasingly become the controlling factor for structural serviceability performance. By accurately predicting crack development during the design phase, it is possible to fully leverage the performance advantages or environmental benefits of new materials while effectively controlling deformation and crack progression during service. Alternatively, based on prediction results, such new material members can be reasonably applied in scenarios with less stringent deformation requirements to ensure compliance with normal serviceability criteria. Based on different fracture mechanisms, traditional methods for analyzing concrete cracks can be divided into two categories. The first category is based on fracture mechanics theory, employing indicators such as the stress intensity factor K, energy release rate G, or strain energy density factor S to assess crack initiation and propagation. Typical applications include the double-K fracture criterion and the double-G fracture criterion proposed by Xu et al. and Zhao et al. [10–13]. The second category is based on the material mechanics theory, using indicators such as maximum tensile stress, maximum tensile strain, or maximum shear stress to evaluate material cracking. Typical applications include the crack resistance calculation methods based on member edge strain, proposed by Zhao et al.and Cheng et al. [14–16]. The first category focuses on the stress state at the crack tip. Its application requires solving the stress and strain fields around the crack tip, thus presenting difficulties for scenarios involving simultaneous propagation of multiple cracks. This category is more suitable for cases dominated by a single large crack, such as large dams in hydraulic structures. In contrast, the second category focuses on the stress state of the member section. It comprehensively reflects the effect of cracks through the average deformation of the cracked region [17]. Consequently, it can be flexibly applied to scenarios with simultaneous propagation of multiple cracks, such as slabs, beams, columns, and shear walls commonly found in building structure design. As a result, it has been widely adopted in building structure design codes worldwide, such as the Chinese code GB/T 50010-2010 [18], the European code Eurocode 2 [19], and the American code ACI 318 [20]. This study focuses on building structures and, therefore, is also based on the materials mechanics approach.

The key parameter in the material mechanics approach for analyzing member cracking is the inelastic influence coefficient of the section resistance moment. Experimental observations and studies on ordinary reinforced concrete members have shown that before reaching the commonly recognized "member cracking" state (i.e., "material cracking" on the member surface), a small number of fine cracks, difficult to detect with the naked eye, appear on their surface [21–23]. This phenomenon results in the calculated tensile stress at the edge of the member's tension zone, derived from measured strain, being greater than the ultimate tensile stress of concrete. This phenomenon is generally considered to be caused by the inelastic deformation of concrete and is characterized by the inelastic influence coefficient of the section resistance moment (denoted as γ). Early research indicated that the inelastic influence coefficient γ of the section resistance moment for ordinary reinforced concrete members is related to material strength, sectional characteristics (section shape, section depth, reinforcement ratio), and loading conditions (load eccentricity) [14–16,24–26]. Current concrete structure design codes only consider the influence of the main factors, i.e., section shape and section depth [18]. Recent studies have shown that for concrete members with new reinforcement materials and high-strength concrete, the influencing factors and their weights on the inelastic influence coefficient γ of the section resistance moment are far more complex than those for ordinary concrete members. For ordinary reinforced concrete incorporating steel fibers, Gao et al.and Li et al. suggested that the value of γ should account for the thickness of the steel fiber reinforcement layer, the aspect ratio of the steel fibers, and the volumetric dosage [1,2]. For cases where FRP bars replace ordinary steel bars, Jia et al. believed that the value of γ should comprehensively consider the influence of concrete strength and FRP reinforcement ratio [3]. For cases where recycled aggregates replace ordinary aggregates, Chen et al. proposed that the value of γ requires additional correction based on the depth of the compression zone [4]. For cases where reactive powder replaces ordinary cement, Zheng et al., Li et al.and Lu et al. argued that only the influence of the reinforcement ratio needs to be considered [5–7]. Overall, the inelastic influence coefficient γ of the section resistance moment used for calculating member cracking internal forces remains a semi-empirical and semi-theoretical parameter based on experimentation. Its value requires corresponding adjustments according to changes in member constituent materials. The focus of existing research has primarily been on its more detailed classification and refined valuation [27–30].

It is worth noting that, inspired by advances in artificial intelligence, some researchers have begun to explore the use of machine learning methods to analyze the complex cracking behavior of concrete members, aiming to eliminate errors arising from data regression bias in semi-empirical parameters. Talpur et al. compared the differences in accuracy of various machine learning models in predicting the mechanical properties of concrete by predicting the strength of concrete confined with sustainable natural FRP compostes [31]. Ma et al. used an interpretable SHAP model to predict the cracking shear strength of reinforced concrete deep beams [32]. Liang et al. employed a Stacking ensemble model to predict the corrosion-induced cracking behavior of reinforced concrete members [33]. These machine learning models have successfully provided more accurate results under corresponding conditions compared to traditional methods. However, in contrast to traditional "mechanism-driven" methods, machine learning methods are inherently "data-driven" approaches. Their prediction results generally lack clear physical meaning, and the input–output process remains in a "black box".

Indeed, although the commonly defined "member cracking" is not the state of the "first crack appearance" in the member (manifested by a inelastic influence coefficient γ of the section resistance moment greater than one) [15], this method has been well applied in engineering practice. This may be attributed to the fact that during the low-load stage, only very few and extremely fine microcracks are produced, and the cohesion between crack surfaces remains effective. Thus, the corresponding experimental load–displacement curve does not exhibit a distinct turning point, and the impact on concrete durability and stiffness is essentially negligible. Only when the load gradually approaches the "cracking load" and the crack width increases significantly (e.g., to 0.04–0.08 mm, as observed by Guo et al. [21]) does the corresponding experimental load–displacement curve show a clear turning point. At this stage, the intrusion of moisture and air becomes significantly easier, and the sectional stiffness is noticeably reduced, leading to macroscopic effects of decreased durability and stiffness due to "cracking". This indicates that "member cracking" is not equivalent to a meso-level change like "material cracking," but rather tends to be a macro-level change process reflected by alterations in member durability and stiffness. Accordingly, the appearance of visible cracks on the member surface or the inflection point in the load–deformation curve is generally adopted as the criterion for determining member cracking in experiments [15]. Based on this, the present study also employs the material mechanics approach focused on the overall sectional stress state, rather than the fracture mechanics method focused on local crack-tip stress conditions. From an energy perspective, the variation in energy before and after member cracking is analyzed based on the macroscopic mechanical behavior of reinforced concrete members. The critical state of member cracking is characterized by the strain energy. On the one hand, this approach avoids the complex classification of the inelastic influence coefficient γ of the section resistance moment based on sectional geometry and material mechanical properties, thereby eliminating the additional empirical coefficients. On the other hand, it also circumvents the lack of the mechanical basis in the prediction process, ensuring that the results possess clear physical significance. Following this strain energy-based method, the cracking loads of plain and reinforced concrete members under uniaxial stress are analyzed. The mechanism whereby the strain within the tensile zone exceeds the material's ultimate tensile strain (i.e., the strain corresponding to the peak tensile stress) at member cracking is explained. The feasibility and accuracy of the proposed method are verified by using existing experimental results.

2. Characterizing Member Cracking Based on Strain Energy Increment

It is generally accepted that a plain concrete member under axial tension cracks along the plane perpendicular to the tensile force when its sectional stress reaches the material's peak stress [21]. The typical graph of the material's stress–strain curve (σ-ε curve) during this process is shown in Figure 1.

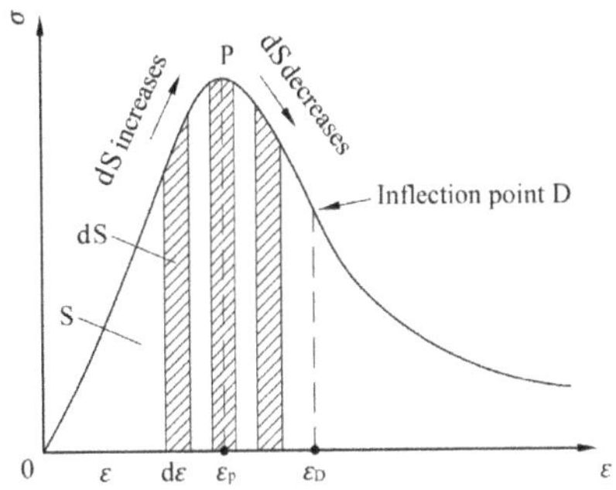

Figure 1. Concrete stress–strain curve under tension.

According to its definition, the strain energy density of the member equals the area S enclosed by the σ-ε curve and the ε-axis. Assuming the member is homogeneous and isotropic, the strain energy U of the member equals the product of the strain energy density S and the member volume V:

$$U(\varepsilon) = V \int_{\varepsilon} \sigma(\varepsilon) \mathrm{d}\varepsilon \tag{1}$$

As illustrated, the variation in the strain energy increment, $\Delta(V \cdot \mathrm{d}S)$, changes from increasing to decreasing near the peak stress point. Owing to the continuity of energy, the cracking point of a plain concrete member under axial tension corresponds to the condition of zero variation in its strain energy increment, i.e., $\Delta(V \cdot \mathrm{d}S) = 0$. This process can be expressed as follows:

$$\frac{\mathrm{d}^2 U(\varepsilon)}{\mathrm{d}\varepsilon^2} = V \frac{\mathrm{d}\sigma(\varepsilon)}{\mathrm{d}\varepsilon} \tag{2}$$

At the cracking of the axially tensioned member, the member strain $\varepsilon_{\mathrm{cr}}$ exactly equals the concrete peak strain ε_{P}, i.e., $\varepsilon_{\mathrm{cr}} = \varepsilon_{\mathrm{P}}$, while the stress–strain curve has a zero slope:

$$\frac{\mathrm{d}\sigma(\varepsilon)}{\mathrm{d}\varepsilon} = 0, \quad \text{when } \varepsilon = \varepsilon_{\mathrm{P}} \tag{3}$$

Consequently, at the cracking of the axially tensioned member, the second-order differential of the strain energy with respect to the member deformation equals zero:

$$\frac{\mathrm{d}^2 U(\varepsilon)}{\mathrm{d}\varepsilon^2} = 0, \quad \text{when } \varepsilon = \varepsilon_{\mathrm{cr}} \tag{4}$$

It should be noted that the derivation is entirely and solely based on the experimental load–deformation curve. Provided that the adopted experimental curve is sufficiently representative, the cracking results predicted by the above equation should also be sufficiently accurate.

The above conclusion also holds true for plain concrete and reinforced concrete members subjected to other loading conditions. Similarly, assume the member is homogeneous and isotropic, and consider the complete process wherein an arbitrary member is subjected to a system of external forces ΣF_i until cracking occurs. Note that the action of a bending moment can also be transformed into the action of a pair of forces (i.e., a force couple). The displacement generated within the member along the direction of each external force F_i is denoted as δ_i, the internal strain field of the member is denoted as ε, and the sum of the strain energies of the concrete and steel is denoted as ΣU_i. According to the principle of work and energy, the work performed by the external forces is transformed into the strain energy of the member; consequently, within any infinitesimal interval prior to cracking, the following relationship always holds:

$$\sum\left[(F_i)^{\mathrm{bf}}\,\mathrm{d}\delta_i\right]=\mathrm{d}\left[\sum U_i(\varepsilon)\right]^{\mathrm{bf}} \tag{5}$$

where the superscript "bf" denotes the state before the member cracks. During this phase, the external force F_i must be continuously increased to induce cracking; therefore, $(F_i)^{\mathrm{bf}}$ continually increases, which in turn causes $\mathrm{d}\left[\sum U_i(\varepsilon)\right]^{\mathrm{bf}}$ to increase as well. Similarly, within the infinitesimal interval immediately preceding the instance of cracking, the following equation applies:

$$\sum\left[(F_i)^{\mathrm{cr}}\,\mathrm{d}\delta_i\right]=\mathrm{d}\left[\sum U_i(\varepsilon)\right]^{\mathrm{cr}} \tag{6}$$

where the superscript "cr" denotes the cracking state of the member. In the infinitesimal interval immediately after cracking, since cracking necessarily involves the formation of new cracks or the propagation of existing ones, this process inevitably results in energy dissipation in forms such as surface energy, acoustic emission, and optical emission (denoted collectively as ΣW_i). Consequently, part of the work done by the external forces is allocated to cover these energy expenditures. Moreover, given that the external force F_i remains constant at the instant of cracking, the following relationship is established:

$$\sum\left[(F_i)^{\mathrm{cr}}\,\mathrm{d}\delta_i\right]=\mathrm{d}\left[\sum U_i(\varepsilon)\right]^{\mathrm{af}}+\mathrm{d}\left(\sum W_i\right) \tag{7}$$

where the superscript "af" denotes the state after the member has cracked. A comparison of Equations (6) and (7) reveals that

$$\mathrm{d}\left[\sum U_i(\varepsilon)\right]^{\mathrm{af}}<\mathrm{d}\left[\sum U_i(\varepsilon)\right]^{\mathrm{cr}} \tag{8}$$

which indicates that $\mathrm{d}\left[\sum U_i(\varepsilon)\right]^{\mathrm{af}}$ decreases in the infinitesimal interval immediately following cracking. In summary, the increment of the member's strain energy, $\mathrm{d}\left[\sum U_i(\varepsilon)\right]$, functions as an increasing quantity before cracking and as a decreasing quantity instantaneously after cracking. At the precise moment of cracking, it transitions from increasing to decreasing. Therefore, the cracking point corresponds to the extremum point of the strain energy increment $\mathrm{d}\left[\sum U_i(\varepsilon)\right]$. That is, the cracking point can also be defined by the condition:

$$\frac{\mathrm{d}^2\left[\sum U_i(\varepsilon)\right]}{\mathrm{d}\varepsilon^2}=0,\quad \text{when } \varepsilon=\varepsilon_{\mathrm{cr}} \tag{9}$$

where the strain energy-strain relationships for the concrete and steel components can still be obtained from the load–deformation curves of the corresponding materials. Since the materials mechanics method generally attributes the material cracking to tensile strain or tensile stress, the load–deformation curves used here should be the experimental curves obtained under pure tension conditions, and the material strain energy should be taken as a function of the normal strain in the tensile direction within the member. When the member is simultaneously subjected to significant shear forces or constraint effects, the load–deformation curves must be derived from new experimental curves that account for the effects of shear or constraints under the equivalent working conditions. This is necessary because these effects alter the material's strain energy response–for instance, shear introduces shear strain energy, while confinement leads to a more complex strain energy composition under triaxial stress states. Theoretically, provided that the adopted load–deformation curve accurately reflects the variation of the strain energy with deformation in the member, the above method can still yield sufficiently accurate prediction results. This aspect will be addressed in future work.

For reinforced and FRP-reinforced concrete members primarily subjected to tensile deformation, the calculation procedure for cracking loads using the above strain energy method can be summarized as follows. Firstly, determine the tensile stress–strain curves of the concrete, steel reinforcement, or FRP reinforcement. Next, based on the sectional geometry, material layout, and material tensile stress–strain relationships, formulate the expression for the total strain energy of the member with the normal strain in the tensile direction serving as the independent variable. Finally, compute the second-order differential of the total strain energy of the member, and take the tensile strain obtained when this second-order differential equals zero as the cracking strain of the member. It is evident that the proposed calculation procedure requires only the fundamental constitutive relationship, i.e., the stress–strain curves of the constituent materials, thereby eliminating the need to introduce the empirical parameter such as the inelastic influence coefficient γ of the section resistance moment through specialized tests. It is important to note that commonly used design codes or manuals generally provide fitted formulas for the stress–strain curves of concrete and steel reinforcement based on experimental data. This allows the stress–strain constitutive relationship of the section materials required by the proposed method to be readily obtained, which will significantly simplify the application of the proposed method.

3. Cracking of Axially Tensioned Members of Steel-Reinforced Concrete

3.1. Case of Conventional Reinforcement Ratio

The deformation analysis of reinforced concrete members under axial tension by Guo et al. [22] indicates that the member does not crack immediately after the member strain ε exceeds the concrete peak strain ε_p. Conversely, the member's bearing capacity N exhibits a short ascending segment before dropping rapidly, forming a small peak on the load–deformation curve (N-ε curve). Cracking of the member occurs only after the bearing capacity reaches this peak. That is, the strain at which the member cracks is greater than the strain at which the concrete cracks. This process is illustrated by curve G in Figure 2.

However, according to the viewpoint that characterizes member cracking by material cracking, the member cracks instantly when the concrete strain reaches the peak strain ε_p. The concrete immediately ceases to carry load, and the tension it carried is transferred to the reinforcement, causing an abrupt increase in the stress and strain of the reinforcement. The bearing capacity of the member remains unchanged throughout this process. It is noted that before cracking, the reinforcement deforms compatibly with the concrete, and the strain in the reinforcement is always equal to the strain in the concrete. After cracking, however, the reinforcement bears the entire tensile force as the concrete ceases to carry load, resulting in a sharp increase in the reinforcement's strain. Compared to the total deformation during the loading process, the deformation accumulated in the reinforcement before cracking is relatively small and can be neglected. Therefore, the incremental deformation of the member at the cracked section can be approximately expressed as:

$$\Delta\varepsilon = \frac{E_c A_c \varepsilon_p}{E_s A_s} \tag{10}$$

where E denotes the elastic modulus, A denotes the cross-sectional area, and the subscripts c and s denote concrete and steel reinforcement, respectively. This process is illustrated by curve M in Figure 2.

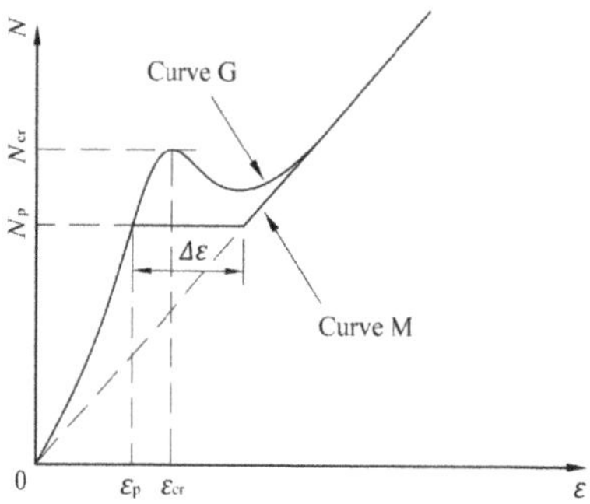

Figure 2. Load–strain curve of axially tensioned member based on classical method and experimental data.

The cracking process revealed by Guo et al. [22] can be reasonably explained by characterizing member cracking through strain energy. Prior to member cracking, the deformations of the concrete and steel reinforcement satisfy the plane-section assumption, and their longitudinal deformations are identical; therefore, we have:

$$\Sigma U_i = \int_0^{\varepsilon} (\sigma_s A_s + \sigma_c A_c) \times (L\,d\varepsilon) \tag{11}$$

where σ denotes stress, and L denotes the member length. At the instance of member cracking, the second-order differential of the strain energy equals zero, yielding:

$$d^2(\Sigma U_i) = \frac{\partial(\sigma_s A_s + \sigma_c A_c)}{\partial\varepsilon} \times (L\,d\varepsilon) = 0 \tag{12}$$

Given that the member length L and the cross-sectional areas A_s and A_c are non-zero, the above equation can be simplified to:

$$\rho\frac{\partial\sigma_s}{\partial\varepsilon} + (1-\rho)\frac{\partial\sigma_c}{\partial\varepsilon} = 0 \tag{13}$$

where ρ represents the reinforcement ratio. Under conventional reinforcement conditions, the strain at member cracking is relatively small, with the steel deformation remaining in the elastic stage; thus the stress–strain relationship obeys Hooke's law, always satisfying:

$$\frac{\partial\sigma_s}{\partial\varepsilon} = E_s \tag{14}$$

For steel reinforcement, E_s is always a positive constant. For concrete, when the strain is less than the peak strain, the stress–strain curve is in the ascending branch, and its slope is always greater than zero; when the strain is greater than the peak strain, the stress–strain curve is in the descending branch, and its slope is always less than zero. Consequently, the positive solution ε_{cr} for Equation (13) can only exist in the region where the strain is greater than the peak strain, which explains the phenomenon that the member cracking strain ε_{cr} exceeds the material cracking strain. Furthermore, Equation (13) also indicates that the member cracking strain ε_{cr} is influenced by the reinforcement ratio ρ, which is consistent with the experimental results reported in references [34,35].

When the reinforcement is uniformly distributed along the width direction of the section (i.e., the section deformation satisfies the plane-section deformation assumption), the axial tensile cracking internal force for any section composed of this reinforced concrete can be solved according to Equation (13) by selecting a set of stress–strain curves for steel and concrete. The tensile stress–strain curve of concrete follows the expression suggested by Guo et al. [21]:

$$\frac{\sigma_c}{f_t} = \begin{cases} 1.2x - 0.2x^6, & x \leq 1 \\ \dfrac{x}{\alpha(x-1)^{1.7} + x}, & x \geq 1 \end{cases} \tag{15}$$

$$x = \frac{\varepsilon}{\varepsilon_p} \tag{16}$$

and the tensile stress–strain curve of steel follows the conventional tri-linear model [22]:

$$\sigma_s = E_s\,\varepsilon \tag{17}$$

Under conventional reinforcement conditions, the cracking strain ε_{cr} for a reinforced concrete axially tensioned member can be determined by substituting Equations (15)–(17) into Equation (13):

$$\rho\frac{E_s}{E_{t,p}} - (1-\rho)\alpha\frac{0.7(x_{cr}-1)^{1.7} + 1.7(x_{cr}-1)^{0.7}}{\left[\alpha(x_{cr}-1)^{1.7} + x_{cr}\right]^2} = 0 \tag{18}$$

where

$$x_{cr} = \frac{\varepsilon_{cr}}{\varepsilon_p} \tag{19}$$

Equations (18) and (19) are applicable to both normal-strength and high-strength concrete. In above equations: α is the parameter for the descending branch of the concrete tensile stress–strain curve, $E_{t,p}$ is the secant modulus of concrete corresponding to the peak tensile stress f_t, and ε_p is the strain of concrete corresponding to the peak tensile stress, given respectively by [22]:

$$\alpha = 0.312 f_t^2 \tag{20}$$

$$E_{t,p} = \frac{(1.45 + 0.628 f_t) \times 10^4}{1.2} \tag{21}$$

$$\varepsilon_p = 65 \times 10^{-6} f_t^{0.54} \tag{22}$$

The corresponding cracking internal force N_{cr} is:

$$N_{cr} = \left[\frac{E_s}{E_{t,p}} x_{cr}\rho + \frac{x_{cr}}{\alpha(x_{cr}-1)^{1.7} + x_{cr}}(1-\rho)\right] f_t A \tag{23}$$

where A represents the cross-sectional area of the member. Calculations were performed using the equations above for conventianally reinforced sections, with f_t ranging from 1.5 to 2.5 N/mm^2, approximately corresponding to concrete strength grades C20 to C50, covering essentially the normal-strength concrete range. The calculation results are shown in Figure 3.

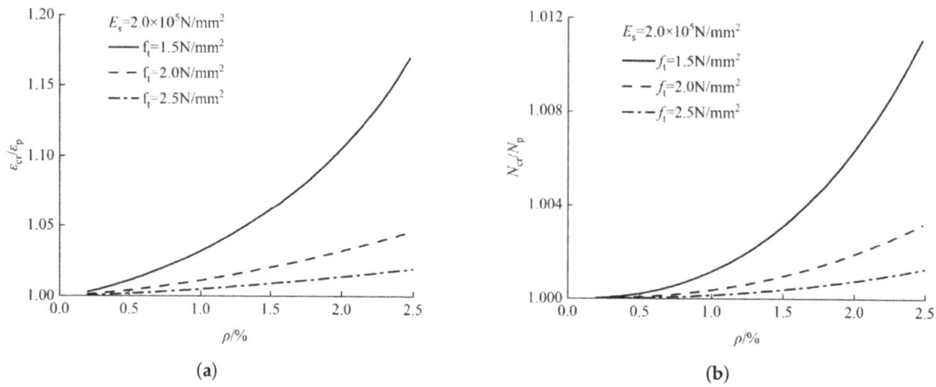

Figure 3. Relative cracking loads and strains of reinforced concrete axially tensioned members. (**a**) Relative strain ratios; (**b**) Relative load ratio.

The calculation results reflect the experimental phenomenon that "the member cracking strain exceeds the concrete cracking strain," indicating the role of reinforcement in this strain enhancement phenomenon. Typically, the contribution of reinforcement to the increase in the cracking load of concrete members is relatively small, generally not exceeding 1%. Thus, calculating the cracking internal force based on the concrete tensile peak strain already provides sufficient accuracy. However, reinforcement can significantly increase the member's cracking strain (e.g., by approximately 5% for concrete grade C30), which is clearly beneficial for delaying member cracking and enhancing crack resistance. Furthermore, this enhancement is particularly evident in members with lower concrete strength grades (e.g., the cracking strain can be increased by 5% to 15% for concrete grade C20).

3.2. Case of High Reinforcement Ratio

Analysis of the critical cracking condition for reinforced concrete members under axial tension, as expressed by Equation (13), indicates that the solution to this equation depends on the relative magnitudes of the slopes of the stress–strain curves for the concrete and steel. Under conventional reinforcement conditions, member cracking occurs shortly after the concrete reaches its peak tensile strain, while the steel remains elastic with a constant stress–strain curve slope. Since the slope of the concrete stress–strain curve reaches its minimum value at the inflection point of the descending branch (point D on the curve in Figure 1), an excessively high reinforcement ratio ρ may cause the left-hand side of Equation (13) to remain greater than zero when calculated using the concrete's inflection point slope and the steel's constant slope, resulting in no positive solution existing for the member during the steel's elastic strain stage. This implies that cracking cannot occur while the steel deforms elastically. It can be inferred that cracking in such members must occur after the steel's elastic deformation stage. At this stage, the concrete strain becomes extremely large, and the slope of its stress–strain curve approaches zero. The positive solution for Equation (13) will then be determined by the slope of the steel's stress–strain curve. That is, a member with a high reinforcement ratio will crack at the point where the slope of the steel's stress–strain curve equals zero. For mild steel, this point corresponds to the yield point of the steel. For hard steel (i.e., steel with no distinct yield point), this point corresponds to the fracture point of the steel. The critical transition state from conventional reinforcement to high reinforcement ratio occurs when the member strain ε reaches the concrete's inflection point strain ε_D precisely at the moment of member cracking. The reinforcement ratio at this state is denoted as the inflection point reinforcement ratio ρ_D. When the reinforcement ratio exceeds ρ_D, the member will maintain macroscopically uncracked behavior during the steel's elastic deformation stage, meaning its stiffness and durability remain essentially uncompromised, and its crack resistance is significantly enhanced.

The concrete's inflection point strain ε_D can be determined by locating the inflection point of the function, i.e., by setting the second derivative of the concrete stress with respect to strain equal to zero. Using the previously selected stress–strain relationships for concrete and steel, the following equation is obtained:

$$0.595x_D\left[\alpha(x_D-1)^{1.7}+x_D\right]-(x_D-1)[1.7+0.7(x_D-1)]\left[1.7\alpha(x_D-1)^{0.7}+1\right]=0 \quad (24)$$

$$\varepsilon_D = x_D\varepsilon_p \quad (25)$$

The inflection point reinforcement ratio ρ_D can be calculated as follows:

$$\rho_D\frac{E_s}{E_{t,p}}-(1-\rho_D)\alpha\frac{0.7(x_D-1)^{1.7}+1.7(x_D-1)^{0.7}}{\left[\alpha(x_D-1)^{1.7}+x_D\right]^2}=0 \quad (26)$$

For common concrete sections, the calculation results are presented in Table 1.

Table 1. Inflection Point Reinforcement Ratio for Common Concrete Sections (%).

E_s ($\times10^5$ N/mm^2)	f_t (N/mm^2)				
	1.0	**1.5**	**2.0**	**2.5**	**3.0**
2.0	1.32	2.95	5.14	7.81	10.87
2.1	1.26	2.82	4.91	7.46	10.40

E_s represents the elastic modulus of steel, f_t represents the peak tensile strength of concrete.

4. Cracking of Flexural Members of Steel- and FRP-Reinforced Concrete

4.1. Energy-Based Criterion for Critical Cracking State

Before the member cracks, the deformation of the flexural member's section satisfies the plane section assumption, and no relative slip occurs between the concrete and the reinforcement. Therefore, the sum of the strain energy of the concrete and steel, ΣU_i, equals the work done by the external moment M over the sectional rotation φ, thus:

$$\sum U_i(\varepsilon)=\int_0^{\varphi} M\,\mathrm{d}\varphi \quad (27)$$

As the member approaches cracking, the development of inelastic deformation at the impending cracked section means the sectional deformation no longer strictly satisfies the plane-section assumption. However, extensive experimental measurements have demonstrated that the average strain along the tensile direction of the member still satisfies the plane-section assumption [22]. It is noted that the tensile stress–strain curve used in the proposed method is also derived from experimental measurements, where the deformation values are averages that inherently account for inelastic deformation. Therefore, the plane-section assumption remains applicable at the critical cracking state. According to Equation (9), the critical condition for flexural cracking is:

$$\frac{\mathrm{d}^2\left[\sum U_i(\varepsilon)\right]}{\mathrm{d}\varepsilon^2}=\frac{\partial M}{\partial\varphi}=0 \quad (28)$$

Consider a double-reinforced section with conventional reinforcement ratio, whose sectional dimensions and stress–strain distribution are shown in Figure 4.

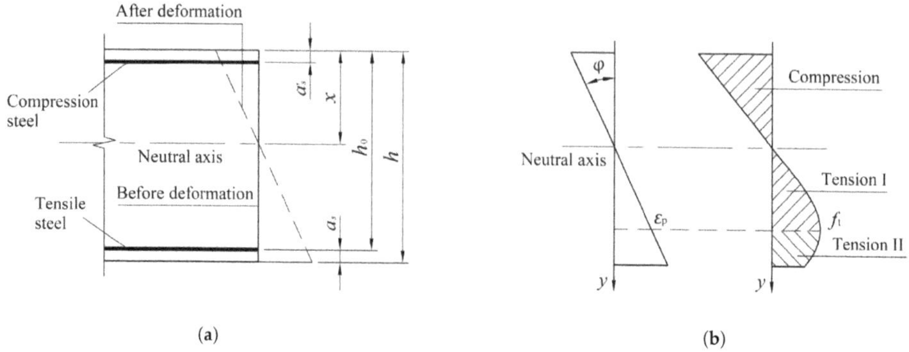

Figure 4. Sectional dimensions and stress–strain distribution of a doubly reinforced flexural member. (**a**) Sectional dimensions; (**b**) Sectional stress-strain distribution.

Since the sectional strain at member cracking is very small, both the compressive and tensile reinforcement remain in the elastic stage before cracking, and the stress–strain relationship of the concrete in the compression zone is essentially linear and can be simplified as a triangular distribution, namely:

$$\sigma_{\mathrm{s}}^{\mathrm{t}} = E_{\mathrm{s}}\varphi(h - x - a_s) \tag{29}$$

$$\sigma_{\mathrm{s}}^{\mathrm{c}} = E_{\mathrm{s}}\varphi(x - a_s') \tag{30}$$

$$\sigma_{\mathrm{c}}^{\mathrm{c}} = E_{\mathrm{c}}\varphi y_{\mathrm{c}} \tag{31}$$

where E_{c} is the secant modulus of concrete in compression, which can be considered slightly less than the initial tangent modulus $E_{\mathrm{c}0}$ during the loading process from zero to member cracking; y_{c} is the distance from the calculation point to the neutral axis when determining the compressive strain of the concrete in the compression zone;the superscripts c and t denote compression and tension, respectively.

The stress distribution of concrete in the tension zone, $\sigma_{\mathrm{c}}^{\mathrm{t}}$, is relatively complex. Since cracking occurs only after the concrete strain exceeds the peak tensile strain, the tension zone is divided into two parts: the region near the neutral axis where $\varepsilon \leq \varepsilon_{\mathrm{p}}$ (Tension Zone I) and the region near the tensile edge where $\varepsilon > \varepsilon_{\mathrm{p}}$ (Tension Zone II). Tension Zone I lies on the ascending branch of the concrete tensile stress–strain curve. According to the expression suggested by Guo et al. [21], the stress in this zone is given by:

$$\sigma_{\mathrm{c}}^{\mathrm{t,I}} = \left[1.2\left(\frac{\varphi y_{\mathrm{t}}}{\varepsilon_{\mathrm{p}}}\right) - 0.2\left(\frac{\varphi y_{\mathrm{t}}}{\varepsilon_{\mathrm{p}}}\right)^6\right] f_{\mathrm{t}} \tag{32}$$

Tension Zone II lies on the descending branch of the concrete tensile stress–strain curve, which exhibits a convex-to-concave shape. The expressions commonly suggested [21,36] are complex and inconvenient for calculation. We propose a straight-line simplification for this curve segment. A straight line drawn through the peak point P and the inflection point D of the curve will inevitably intersect the concave segment at a point E, as shown in Figure 5.

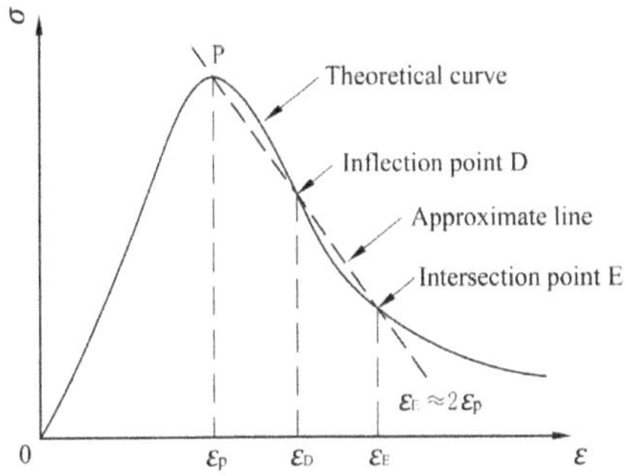

Figure 5. Schematic of the approximate line passing through the descending branch of the concrete stress–strain curve.

Analysis of existing experimental data and trial calculations reveals that the tensile strain at the member edge at cracking generally lies near point E. It is noted that within the strain range of Tension Zone II, the curvature of the original stress–strain curve is small, resulting in minor error when replaced by a straight line. Furthermore, the approximating line value is less than the curve value before the inflection point and greater than the curve value after the inflection point; averaging these two deviations can further reduce the error. Therefore, this straight line can be used to effectively approximate the descending branch of the concrete stress–strain curve. Applying normalization to the concrete stress–strain curve, with the ratio of strain to peak tensile strain as the x-axis and the ratio of stress to peak tensile stress as the y-axis, the shape of the descending branch depends solely on the concrete tensile peak stress f_t [21]. In this dimensionless coordinate system, the approximating line always passes through the point (1,1), i.e., the peak stress point. Its slope k depends on the curve shape, meaning the slope of the approximating line is also determined by the concrete tensile peak stress f_{t}. For the common range of f_{t} from 0.8 to 3.2 MPa, the k-f_{t} relationship is well approximated by a quadratic function. Fitting with a quadratic term yields:

$$k = -\left(0.0252 f_{\mathrm{t}}^2 + 0.1728 f_{\mathrm{t}} - 0.0752\right) \quad (33)$$

Comparing the slope estimated by the above equation with the slope calculated from the theoretical stress–strain curve, the fitted formula has a root mean square error of 0.001903 and a coefficient of determination of 0.9999, meeting the required accuracy. Therefore, the concrete stress in Tension Zone II can be simplified as:

$$\sigma_{\mathrm{c}}^{\mathrm{t,II}} = \left[k\left(\frac{\varphi y_{\mathrm{t}}}{\varepsilon_{\mathrm{p}}} - 1\right) + 1\right] f_{\mathrm{t}} \quad (34)$$

Base on the force equilibrium conditions of the section, the equations for the forces are obtained as follows:

$$N_{\mathrm{s}}^{\mathrm{c}} + N_{\mathrm{c}}^{\mathrm{c}} = N_{\mathrm{s}}^{\mathrm{t}} + N_{\mathrm{c}}^{\mathrm{t,I}} + N_{\mathrm{c}}^{\mathrm{t,II}} \quad (35)$$

$$N_{\mathrm{s}}^{\mathrm{t}} = A_{\mathrm{s}} E_{\mathrm{s}} \varphi (h - x - a_{\mathrm{s}}) \quad (36)$$

$$N_{\mathrm{s}}^{\mathrm{c}} = A_{\mathrm{s}}' E_{\mathrm{s}} \varphi (x - a_{\mathrm{s}}') \quad (37)$$

$$N_{\mathrm{c}}^{\mathrm{c}} = \int_0^x E_{\mathrm{c}} \varphi y_{\mathrm{c}} b \mathrm{d} y_{\mathrm{c}} = \frac{1}{2} b E_{\mathrm{c}} \varphi x^2 \quad (38)$$

$$N_c^{t,I} = \int_0^{\frac{\varepsilon_p}{\varphi}} \left[1.2\left(\frac{\varphi y_t}{\varepsilon_p}\right) - 0.2\left(\frac{\varphi y_t}{\varepsilon_p}\right)^6\right] f_t b \mathrm{d}y_t = \frac{4}{7} f_t b \left(\frac{\varepsilon_p}{\varphi}\right) \tag{39}$$

$$N_c^{t,II} = \int_{\frac{\varepsilon_p}{\varphi}}^{h-x} \left[k\left(\frac{\varphi y_t}{\varepsilon_p} - 1\right) + 1\right] f_t b \mathrm{d}y_t = f_t b\left[\frac{k}{2}(h-x)^2\left(\frac{\varphi}{\varepsilon_p}\right) + (1-k)(h-x) + \left(\frac{k}{2} - 1\right)\left(\frac{\varepsilon_p}{\varphi}\right)\right] \tag{40}$$

Similarly, base on the moment equilibrium condition of the section, the equation for the bending moment is obtained as follows:

$$M = M_s^t + M_s^c + M_c^c + M_c^{t,I} + M_c^{t,II} \tag{41}$$

$$M_s^t = A_s E_s \varphi (h - x - a_s)^2 \tag{42}$$

$$M_s^c = A_s' E_s \varphi (x - a_s')^2 \tag{43}$$

$$M_c^c = \int_0^x y_c \times (E_c \varphi y_c b \mathrm{d}y_c) = \frac{1}{3} b E_c \varphi x^3 \tag{44}$$

$$M_c^{t,I} = \int_0^{\frac{\varepsilon_p}{\varphi}} y_t \times \left\{\left[1.2\left(\frac{\varphi y_t}{\varepsilon_p}\right) - 0.2\left(\frac{\varphi y_t}{\varepsilon_p}\right)^6\right] f_t b \mathrm{d}y_t\right\} = \frac{3}{8} f_t b \left(\frac{\varepsilon_p}{\varphi}\right)^2 \tag{45}$$

$$M_c^{t,II} = \int_{\frac{\varepsilon_p}{\varphi}}^{h-x} y_t \times \left\{\left[k\left(\frac{\varphi y_t}{\varepsilon_p} - 1\right) + 1\right] f_t b \mathrm{d}y_t\right\} = f_t b\left[\frac{k}{3}(h-x)^3\left(\frac{\varphi}{\varepsilon_p}\right) + \frac{1-k}{2}(h-x)^2 + \left(\frac{k}{6} - \frac{1}{2}\right)\left(\frac{\varepsilon_p}{\varphi}\right)^2\right] \tag{46}$$

Note that $f_t = E_{t,p}\varepsilon_p$, where $E_{t,p}$ is the secant modulus of concrete at the peak tensile stress. The $E_{t,p}$ and the initial tensile modulus of concrete E_t (the secant modulus corresponding to $0.5f_t$) satisfy the relationship [22]: $E_t = 1.2E_{t,p}$, and $E_t = (1.45 + 0.628f_t) \times 10^4$. According to the analysis by Guo et al. [22], the initial tensile modulus of concrete E_t is slightly less than the initial tangent modulus of concrete in compression E_{c0}. Furthermore, E_c during the loading process from zero to cracking can always be considered slightly less than E_{c0}. Therefore, it is approximately valid that $E_c = E_t$ and $E_c = 1.2E_{t,p}$. Consequently, the system of force Equations (35)–(40) can be simplified to:

$$\begin{aligned} &\left(\frac{3}{5} - \frac{k}{2}\right)\left(\frac{\varphi}{\varepsilon_p}\right)x^2 + \left[1.2\frac{E_s}{E_t}\left(\frac{\varphi}{\varepsilon_p}\right)\left(\frac{A_s}{b} + \frac{A_s'}{b}\right) + hk\left(\frac{\varphi}{\varepsilon_p}\right) - k + 1\right]x - \\ &\quad 1.2\frac{E_s}{E_t}\left(\frac{\varphi}{\varepsilon_p}\right)\left[\frac{A_s'}{b}a_s' + \frac{A_s}{b}(h - a_s)\right] - \frac{1}{2}kh^2\left(\frac{\varphi}{\varepsilon_p}\right) + \left(\frac{3}{7} - \frac{k}{2}\right)\left(\frac{\varepsilon_p}{\varphi}\right) + h(k-1) = 0 \end{aligned} \tag{47}$$

The sectional bending moment M can be simplified to:

$$\begin{aligned} \frac{M}{f_t b} = &\left[\frac{k}{3}(h-x)^3 + \frac{2}{5}x^3\right]\left(\frac{\varphi}{\varepsilon_p}\right) + 1.2\frac{E_s}{E_t}\left[A_s'(x - a_s')^2 + A_s(h - x - a_s)^2\right]\left(\frac{\varphi}{\varepsilon_p}\right) + \\ &\left(\frac{k}{6} - \frac{1}{8}\right)\left(\frac{\varepsilon_p}{\varphi}\right)^2 + \frac{1-k}{2}(h-x)^2 \end{aligned} \tag{48}$$

Equation (47) is a quadratic equation in terms of the compression zone depth x with respect to the deformation φ. From this equation, the positive solution for x as a function of φ, denoted $x = x(\varphi)$, can be obtained. Substituting this into Equation (48), the expression for the sectional bending moment M as a function of the deformation φ, denoted $M = M(\varphi)$, can be derived. Finally, the critical deformation at member cracking, φ_{cr}, can be determined from Equation (28):

$$\frac{\partial M}{\partial \varphi} = 0, \quad \text{when } \varphi = \varphi_{cr} \tag{49}$$

4.2. Model Validation with Experimental Results

For flexural members of steel- and FRP-reinforced concrete with moderate reinforcement spacing and reasonable sectional geometry (i.e., deformations satisfying the plane-section assumption), the critical deformation φ_{cr} at member cracking and the corresponding depth x_{cr} of the compression zone can be calculated using Equations (47) and (48). Substituting φ_{cr} into Equation (48) yields the cracking moment M_{cr} of the member. To validate the effectiveness and predictive accuracy of the proposed method, four sets of experimental data on the cracking resistance of steel- or FRP-reinforced concrete beams are compared. The data are from tests conducted by Fang et al. [35], Jiang et al. [36], Sun et al. [37], and Zheng et al. [5], respectively. The experimentally measured values are compared with the calculated values from the original authors and the proposed method. These four comparative datasets cover a range from normal-strength to high-strength concrete and from steel to FRP reinforcement, which makes the data broadly representative.

4.2.1. Effect of Concrete Strength: Normal- to High-Strength

Fang et al. designed and tested 12 steel-reinforced concrete beams and additionally collected test data for 18 beams from the literature [35]. The cross-sectional widths of the beams ranged from 100 to 200 mm, and the heights ranged from 150 to 300 mm. The cube compressive strength f_{cu} of concrete ranged from 20.0 to 82.5 N/mm^2, covering both normal- and high-strength concrete. The reinforcement consisted of ordinary steel bars, with a reinforcement ratio ranging from 0.64% to 1.60%. Fang et al. proposed that the axial tensile strength f_{t} and the elastic modulus E_{c} of concrete be converted according to the following formulas [38]:

$$f_{\mathrm{t}} = 0.395 f_{\mathrm{cu}}^{0.55} \tag{50}$$

and

$$E_{\mathrm{c}} = \left(1.45 + 0.628 f_{\mathrm{t}}\right) \times 10^4 \tag{51}$$

Fang et al. primarily considered the influence of concrete strength on the cracking moment. Their calculation formula reads:

$$M_{\mathrm{cr}} = \gamma_{\mathrm{F}} W_0 f_{\mathrm{t}} \tag{52}$$

where γ_{F} is the inelastic influence coefficient of the section resistance moment, which is adjusted for concrete strength (referred to as the calculated inelastic influence coefficient by Fang et al. [35]):

$$\gamma_{\mathrm{F}} = \begin{cases} 1.55, & f_{\mathrm{cu}} < 30\ \mathrm{N/mm^2} \\ 1.4, & 30\ \mathrm{N/mm^2} \le f_{\mathrm{cu}} \le 60\ \mathrm{N/mm^2} \\ 1.1, & f_{\mathrm{cu}} > 60\ \mathrm{N/mm^2} \end{cases} \tag{53}$$

and W_0 is the elastic section modulus of the transformed section at the tensile edge.

A comparison between the calculated cracking moments from the Fang method and the proposed method with the experimentally measured values is shown in Figure 6. The Fang method yielded an average error of 0.57 kN·m and a root mean square error of 0.75 kN·m, while the proposed method yielded an average error of 0.42 kN·m and a root mean square error of 0.87 kN·m. Comparing the ratio of calculated values to measured values, the Fang method had an average value of 0.96 and a coefficient of variation of 0.10, while the proposed method had an average value of 1.07 and a coefficient of variation of 0.13. It can be concluded that for both normal- and high-strength concrete, the proposed method agrees well with the experimental results, and its prediction accuracy is essentially equivalent to that of the Fang method. It is important to note that the proposed method requires only the fundamental constitutive relationship of the material, i.e., the material's tensile stress–strain curve. In contrast to the Fang method, the proposed method eliminates the need to introduce a specially adjusted inelastic influence coefficient γ_F of the section resistance moment based on concrete strength, making it more straightforward for practical application.

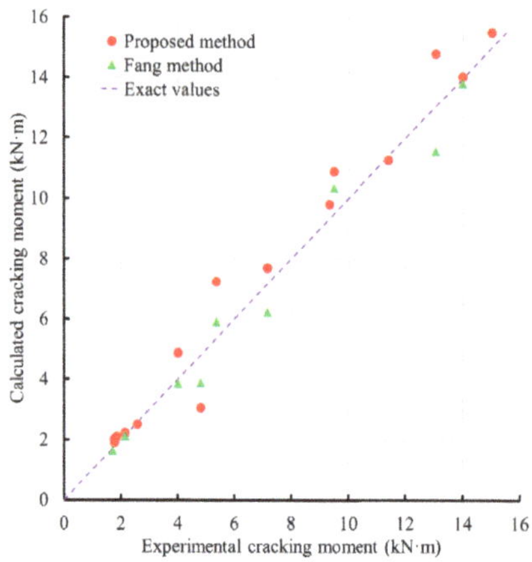

Figure 6. Comparison of measured and calculated cracking moments of normal- and high-strength concrete beams.

4.2.2. Effect of Concrete Strength: High-Strength

Jiang et al. designed and tested 11 steel-reinforced concrete beams [36]. The cross-sectional width of the beams was 180 mm, and the height was 210 mm. The concrete was high-strength C70 grade, with a characteristic axial compressive strength of 51.2 N/mm^2 and a characteristic axial tensile strength of 3.63 N/mm^2, both at a 95% assurance rate. The elastic modulus of the concrete E_c was 38.2 kN/mm^2. The reinforcement consisted of two groups of ordinary steel bars with yield strengths of 560.8 N/mm^2 and 429.1 N/mm^2, respectively. The elastic moduli of the steel bars were 201.1 kN/mm^2 and 199.1 kN/mm^2, respectively, and the reinforcement ratio ranged from 0.70% to 1.40%. According to Ref. [39], the coefficient of variation δ_c for the concrete tensile and compressive strengths is taken as 0.1. Therefore, the mean axial tensile strength and mean axial compressive strength used for checking the test data are given by the following formulas [18,39]:

$$f_m = \frac{f_k}{1 - 1.645\,\delta_c} \quad (54)$$

where f denotes the tensile or compressive strength, the subscript m denotes the mean value and the subscript k denotes the characteristic value at a 95% assurance rate. Thus, the axial tensile strength f_t and axial compressive strength f_c used for verification are calculated as 4.34 N/mm^2 and 61.3 N/mm^2, respectively.

Similarly, Jiang et al. primarily considered the influence of concrete strength on the cracking moment. Their calculation formula reads:

$$M_{cr} = \gamma_J\, W_0\, f_t \quad (55)$$

where γ_J is the inelastic influence coefficient of the section resistance moment, which is modified by the regression based on the concrete strength grade:

$$\gamma_J = 1.75 \times \left(1.1 - 0.015\, f_{cu}^{2/3}\right) \quad (56)$$

where f_{cu} is the cube compressive strength of concrete, which can be converted according to Ref. [38]:

$$f_{cu} = \frac{f_c}{\alpha_{c1}} \tag{57}$$

where α_{c1} is the ratio of the axial compressive strength to the cube compressive strength of concrete, taken as 0.76 for C50 and below, 0.82 for C80, with linear interpolation for intermediate values. Thus, the cube compressive strength f_{cu} of concrete is calculated as 76.6 N/mm^2.

A comparison between the calculated cracking moments from the Jiang method and the proposed method against the experimentally measured values is provided in Table 2. The Jiang method yielded a mean error of 0.19 kN·m and a root mean square error of 0.49 kN·m, whereas the proposed method yielded a mean error of −0.34 kN·m and a root mean square error of 0.56 kN·m. Comparing the ratio of calculated values to measured values, the Jiang method had an average value of 1.02 and a coefficient of variation of 0.05, whereas the proposed method had an average value of 0.96 and a coefficient of variation of 0.05. It can be concluded that for high-strength concrete, the proposed method agrees well with the experimental results and achieves prediction accuracy very close to that of the Jiang method. Similarly, it is important to note that the proposed method requires only the fundamental constitutive relationship, i.e., the material's tensile stress–strain curve, while the Jiang method requires the introduction of a inelastic influence coefficient γ_J of the section resistance moment obtained from specialized tests, making the proposed method more straightforward for practical application.

Table 2. Cracking moments for high-strength concrete beams: measured vs. calculated.

$b \times h$ (mm)	h_0 (mm)	A_s (mm^2)	E_s (kN/mm^2)	M_{exp} (kN·m)	M_{Pro} (kN·m)	M_{Jiang} (kN·m)	$\frac{M_{Pro}}{M_{exp}}$	$\frac{M_{Jiang}}{M_{exp}}$
179 × 207	171	226	201.1	7.88	7.71	8.38	0.98	1.06
181 × 213	177	339	201.1	8.51	8.59	9.15	1.01	1.07
180 × 205	169	452	201.1	8.19	8.25	8.59	1.01	1.05
182 × 208	172	226	201.1	7.68	7.91	8.60	1.03	1.12
179 × 212	176	339	201.1	9.75	8.42	8.97	0.86	0.92
181 × 209	173	452	201.1	9.40	8.61	8.97	0.92	0.95
181 × 209	173	226	201.1	8.51	7.94	8.64	0.93	1.02
181 × 214	178	339	201.1	9.04	8.67	9.23	0.96	1.02
179 × 207	171	452	201.1	8.59	8.37	8.71	0.97	1.01
180 × 208	172	226	199.1	8.49	7.82	8.51	0.92	1.00
179 × 211	175	452	199.1	8.70	8.67	9.05	1.00	1.04

b and h represent the section width and height, respectively; h_0 represents the effective depth of the section; A_s represents the area of reinforcement; E_s represents the elastic modulus of steel; M_{exp} is the experimentally measured value; M_{Pro} and M_{Jiang} are the calculated values from the proposed method and the Jiang method, respectively.

4.2.3. Effect of Reinforcement Ratio: Steel Reinforcement in High-Strength Concrete

Sun et al. designed and tested eight steel-reinforced concrete beams [37]. The cross-sectional width of the beams was 120 mm, and the height was 250 mm. The concrete was high-strength concrete modified with reactive powder and steel fibers, with a cube compressive strength f_{cu} of 120 N/mm^2, an axial tensile strength f_t of 6.9 N/mm^2, and an elastic modulus E_c of 45.3 kN/mm^2. The reinforcement consisted of high-strength steel bars, with measured yield strengths ranging from 531 to 570 N/mm^2, an elastic modulus of 200.0 kN/mm^2, and a reinforcement ratio ranging from 0.87% to 16.35%.

Sun et al. primarily considered the influence of the reinforcement ratio on the cracking moment under conditions of high-strength concrete and high-strength steel reinforcement. Their proposed calculation formula reads:

$$M_{cr} = \gamma_S W_0 f_t \quad (58)$$

where γ_S is the inelastic influence coefficient of the section resistance moment, which is adjusted for the reinforcement ratio:

$$\gamma_S = \begin{cases} 1.33 + 12\rho, & \rho \leq 3.98\% \\ 1.81, & \rho > 3.98\% \end{cases} \quad (59)$$

where ρ is the reinforcement ratio.

A comparison of the calculated cracking moments from the Sun method and the proposed method against the experimentally measured values is shown in Figure 7. For the proposed method, the tensile stress–strain relationship for steel-fiber-reinforced concrete was still adopted from Equations (32)–(34) for ordinary concrete. The results indicate that the Sun method yielded a mean error of −1.41 kN·m and a root mean square error of 1.89 kN·m, whereas the proposed method yielded a mean error of −0.23 kN·m and a root mean square error of 1.59 kN·m. For the ratio of calculated values to measured values, the Sun method had an average value of 0.92 and a coefficient of variation of 0.09, whereas the proposed method had an average value of 0.99 and a coefficient of variation of 0.12. It is evident that for high-strength concrete modified with reactive powder and steel fibers under varying reinforcement ratios, the proposed method agrees well with the experimental results and achieves prediction accuracy essentially equivalent to that of the Sun method. Similarly, the proposed method requires only the fundamental constitutive relationship, i.e., the material's tensile stress–strain curve. In contrast to the Sun method, which requires correcting the inelastic influence coefficient γ_S of the section resistance moment based on the reinforcement ratio through specialized tests, the proposed method is more straightforward for practical application.

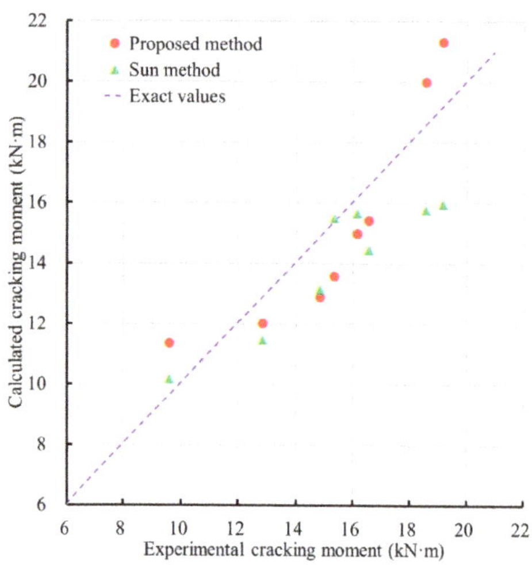

Figure 7. Comparison of measured and calculated cracking moments of high-strength concrete beams modified with reactive powder and steel fibers.

4.2.4. Effect of Reinforcement Ratio: FRP Reinforcement in High-Strength Concrete

Zheng et al. designed and tested eight GFRP-reinforced concrete beams. The cross-sectional width of the beams was 150 mm, and the height was 280 mm. The concrete was high-strength concrete modified with reactive powder and steel fibers. The compressive strength f_c of the 100 mm × 100 mm × 300 mm prism was 102.28 N/mm^2, the axial tensile strength f_t was 10.19 N/mm^2, and the tensile elastic modulus E_c of the concrete was 48.1 kN/mm^2. The GFRP bars had diameters of 5.5 mm, 12 mm, and 14 mm, with tensile strengths of 1159 N/mm^2, 990 N/mm^2, and 836 N/mm^2, respectively. The elastic moduli of the GFRP bars were 49.4 kN/mm^2, 47.6 kN/mm^2, and 50.0 kN/mm^2, respectively. The reinforcement ratio ranged from 0.13% to 2.69%. Two beams with a reinforcement ratio below the minimum longitudinal reinforcement ratio were excluded. Six beams with reinforcement ratios ranging from 0.61% to 2.69% were selected for validation.

Zheng et al. primarily considered the influence of the reinforcement ratio on the cracking moment of beams under conditions of high-strength concrete and FRP reinforcement. Their proposed calculation formula reads:

$$M_{\mathrm{cr}} = \gamma_{\mathrm{Z}} W_0 f_{\mathrm{t}} \tag{60}$$

where γ_{Z} is the inelastic influence coefficient of the section resistance moment, which is adjusted for the reinforcement ratio:

$$\gamma_{\mathrm{Z}} = 1.1 + 6\rho \tag{61}$$

where ρ is the reinforcement ratio.

A comparison of the calculated cracking moments from the Zheng method and the proposed method against the experimentally measured values is provided in Table 3. For the proposed method, the tensile stress–strain relationship for steel-fiber-reinforced concrete was also adopted from Equations (32)–(34) for ordinary concrete. The results indicate that the Zheng method yielded a mean error of −1.00 kN·m and a root mean square error of 1.32 kN·m, whereas the proposed method yielded a mean error of −1.00 kN·m and a root mean square error of 1.63 kN·m. For the ratio of calculated values to measured values, the Zheng method had an average value of 0.96 and a coefficient of variation of 0.04, whereas the proposed method had an average value of 0.96 and a coefficient of variation of 0.05. It is evident that for high-strength GFRP-reinforced concrete modified with reactive powder and steel fibers under varying reinforcement ratios, the proposed method agrees well with the experimental results and achieves prediction accuracy very close to that of the Zheng method. Similarly, the proposed method requires only the fundamental constitutive relationship, i.e., the material's tensile stress–strain curve. Again, in contrast to the Zheng method, which requires correcting the inelastic influence coefficient γ_{Z} of the section resistance moment based on the reinforcement ratio through specialized tests, the proposed method is more straightforward for practical application.

Table 3. Cracking moments for GFRP-reinforced concrete beams: measured vs. calculated.

$b \times h$ (mm)	h_0 (mm)	A_s (mm^2)	E_s (kN/mm^2)	M_{exp} (kN·m)	M_{Pro} (kN·m)	M_{Zheng} (kN·m)	$\frac{M_{\mathrm{Pro}}}{M_{\mathrm{exp}}}$	$\frac{M_{\mathrm{Zheng}}}{M_{\mathrm{exp}}}$
150 × 280	249	226	47.6	24.00	24.12	22.91	1.00	0.95
150 × 280	230.5	452	47.6	24.00	24.30	23.85	1.01	0.99
150 × 280	230.5	565	47.6	24.00	24.44	24.33	1.02	1.01

Table 3. *Cont.*

$b \times h$ (mm)	h_0 (mm)	A_s (mm^2)	E_s (kN/mm^2)	M_{exp} (kN·m)	M_{Pro} (kN·m)	M_{Zheng} (kN·m)	$\frac{M_{Pro}}{M_{exp}}$	$\frac{M_{Zheng}}{M_{exp}}$
150 × 280	230.5	678	47.6	27.00	24.58	24.81	0.91	0.92
150 × 280	228.5	769	50.0	27.00	24.70	25.22	0.91	0.93
150 × 280	228.5	923	50.0	27.00	24.89	25.89	0.92	0.96

b and h represent the section width and height, respectively; h_0 represents the effective depth of the section; A_s represents the area of GFRP reinforcement; E_s represents the elastic modulus of GFRP; M_{exp} is the experimentally measured value; M_{Pro} and M_{Zheng} are the calculated values from the proposed method and the Zheng method, respectively.

4.2.5. Discussion of Results

As mentioned above, the four sets of beam tests encompass normal-strength and high-strength concrete, as well as ordinary steel bars, high-strength steel bars, and FRP reinforcement. The proposed strain energy-based method demonstrates good agreement with the experimentally measured values across all these conditions. Comparing the ratio of the measured value to the calculated value, for all four sets of test data, the average ratio is 1.01, with a coefficient of variation of 0.12. These results indicate that the calculated results from the proposed method are generally very close to the experimental results, although the coefficient of variation is slightly large. This discrepancy arises from two main reasons. On the one hand, the experimental data for concrete cracking inherently exhibit significant scatter, with the coefficient of variation ranging from 0.16 to 0.10 as the concrete strength grade varies from C25 to C80 [39]. On the other hand, the axial tensile strength of concrete, f_t, which significantly influences the calculation of the cracking moment, is partly derived, potentially amplifying the error. Overall, the deviation between the calculated values and experimental values remains within an acceptable range. The overall accuracy is satisfactory. This demonstrates the feasibility of the strain energy-based method for calculating the cracking loads of members.

It is worth noting that the calculation error of the proposed method is significantly larger for the following two cases than that observed in the aforementioned four sets of tests: the ultra-high-strength reinforced concrete beams with the concrete cube compressive strength exceeding 150 N/mm^2 tested by Wang et al. [29], and the recycled aggregate concrete beams using construction waste tested by Kong et al. [8]. For the former case (Wang et al.), the discrepancy arises because the concrete in the tensile zone remains in the elastic deformation stage without entering the inelastic deformation stage at cracking for ultra-high-strength concrete beams [29]. Consequently, the stress–strain distribution at the cracked section does not satisfy the assumptions of the proposed method, as illustrated in Figure 4, leading to considerable deviation in the calculated results. This indicates that the proposed method is not suitable for scenarios involving brittle fracture, such as cracking in ultra-high-strength concrete beams. For the latter case (Kong et al.), the increased error may be attributed to the numerous initial microcracks in the recycled aggregate compared to conventional natural aggregate, which reduce its deformation resistance. Consequently, the deformation characteristics under load differ significantly from those of conventional aggregate concrete. Thus, noticeable deviations occur in the calculated results when the proposed method still employs the stress–strain relationship based on conventional concrete. This indicates that the proposed method must select an appropriate stress–strain relationship according to the properties of the constituent materials. Specifically, the existing constitutive relationships are applicable only to normal-strength and high-strength concrete incorporating conventional natural aggregates or modified with steel fibers or reactive powder.

4.3. Model Assessment via Inelastic Deformation Analysis

Based on the critical deformation φ_{cr} and the corresponding depth x_{cr} of the compression zone at member cracking obtained from the proposed strain energy method, the ultimate strain of concrete at the tensile edge is given by:

$$\varepsilon_{cr}^{t} = \varphi_{cr}(h - x_{cr}) \tag{62}$$

The ratio of the ultimate strain at the tensile concrete edge to the concrete peak tensile strain, $\varepsilon_{cr}^{t}/\varepsilon_{p}$, reflects the development of sectional inelastic deformation and is closely related to the calculation of the cracking moment. Classical theories for cracking of concrete members [22,34] uniformly adopt a value of 2.0 for this ratio. Fang et al. [35] suggested that this ratio decreases with increasing concrete strength, recommending values of 1.5 for normal-strength concrete and 1.1 for high-strength concrete. In fact, the ultimate strain of the concrete at the tensile edge is influenced by concrete strength, reinforcement quantity and arrangement [22,34,35]. Neglecting these factors or considering only the concrete strength may be incomplete. A singly reinforced rectangular beam section, typical in laboratory studies, is selected, with dimensions $b \times h = 150 \times 300$ mm, the effective depth $h_0 = 270$ mm, and reinforcement of HRB400. The strain ratio is calculated for: high reinforcement ratio ($\rho = 1.40\%$), economical ratios ($\rho = 1.13\%$ and 0.89%), low ratio ($\rho = 0.50\%$), and the minimum ratio ($\rho = 0.35\%$). The results are presented in Table 4.

Table 4. Ratio of concrete ultimate strain at tensile edge to peak tensile strain $\varepsilon_{cr}^{t}/\varepsilon_{p}$.

ρ	f_t (C20~C30)									f_t (C30~C50)					f_t (C50~C80)		
	2.0	**2.2**	**2.4**	**2.5**	**2.6**	**2.7**	**2.8**	**2.9**	**3.0**	**3.1**	**3.2**	**3.3**	**3.4**	**3.5**	**3.7**	**3.9**	**4.1**
1.40%	8.82	4.96	3.62	3.24	2.95	2.73	2.55	2.40	2.28	2.17	2.08	2.00	1.94	1.88	1.78	1.69	1.63
1.13%	4.95	3.60	2.93	2.71	2.53	2.39	2.27	2.16	2.08	2.00	1.93	1.87	1.82	1.77	1.69	1.62	1.57
0.89%	3.59	2.92	2.53	2.38	2.26	2.16	2.07	2.00	1.93	1.87	1.82	1.77	1.73	1.69	1.62	1.57	1.52
0.50%	2.55	2.28	2.09	2.01	1.94	1.88	1.83	1.78	1.74	1.70	1.66	1.63	1.60	1.57	1.53	1.48	1.45
0.35%	2.31	2.11	1.96	1.90	1.85	1.80	1.75	1.71	1.68	1.64	1.61	–	–	–	–	–	–

ρ represents the reinforcement ratio; f_t represents the axial tensile stress of concrete (N/mm^2); the entries marked with "–" indicate sections where under-reinforced failure occurred and thus are omitted.

Classical cracking theories for concrete members are primarily based on experimental results from normal-strength concrete members, roughly corresponding to concrete grades C30 to C50. The average strain ratio $\varepsilon_{cr}^{t}/\varepsilon_{p}$ within this range is analyzed: 2.09 for the economical ratio group, 1.98 for the group combining economical and lower ratios, and 2.04 for the full set of reinforcement ratios. The calculated results are in close agreement with the classical theory. Furthermore, the results demonstrate that the strain ratio decreases with increasing concrete strength and increases with the quantity of reinforcement. This trend aligns with the experimental observations that "concrete becomes more brittle with increasing strength" and that "reinforcement enhances the crack resistance of concrete".

Furthermore, classical cracking theories for concrete members also consider the influence of section height on the development of inelastic deformation. By introducing the section height h into the calculation formula of the inelastic influence coefficient γ of the section resistance moment, these theories reflect the relationship between the member size and the concrete cracking resistance. For rectangular sections, a typical fitted formula for the inelastic influence coefficient γ is given by [22]:

$$\gamma = 1.75 \times \left(0.7 + \frac{120}{h}\right) \tag{63}$$

where h is in millimeters.

In the proposed method, the influence of member size on the concrete cracking load is implicitly reflected through the coupling effect of the plane-section assumption and the concrete stress–strain relationship. For members with different section heights, the strain distribution characteristics at the extreme tensile fiber vary with size according to the plane-section assumption. Larger section heights lead to a more pronounced extent by which the strain in the edge region exceeds the strain corresponding to the peak stress of the concrete. The post-peak characteristics of the concrete stress–strain relationship (i.e., the stress decay law after the peak strain) translate this size-dependent strain variation into differences in stress. These stress differences are subsequently reflected in the calculation of the total strain energy of the member, thereby ultimately manifesting as the influence of the size effect.

For the common range of concrete strength grades, the cracking moments of reinforced concrete beams with different section heights are calculated using the proposed method. The corresponding inelastic influence coefficient γ is then back-calculated. The curve of γ vs. the section height h, obtained from the proposed method, is compared with the typical fitted formula curve in Figure 8. Owing to the significant scatter inherent in the experimental data for this parameter [22], the results calculated by the proposed method also show a relatively wide distribution. Nevertheless, the variation trend aligns fundamentally with the fitted formula results. Moreover, the proposed method similarly captures the influence of member size on the cracking resistance of concrete beams.

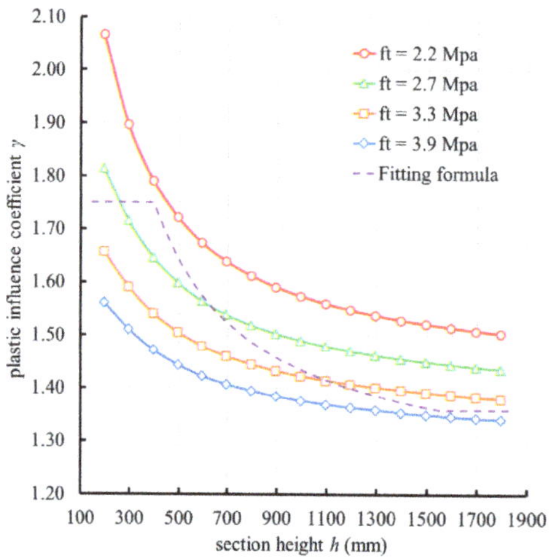

Figure 8. Variation of inelastic influence coefficient γ with section height h.

4.4. Sensitivity Analysis of Key Influencing Parameters

Based on the calculation procedure of the proposed method, the main factors influencing the cracking moment of flexural members are the sectional dimensions, effective depth, concrete strength grade, and the reinforcement ratio of steel or other reinforcement. The effective depth primarily reflects the influence of the concrete cover thickness and the arrangement of the reinforcement. The concrete strength grade primarily reflects the influence of the concrete tensile strength, tensile elastic modulus, and tensile stress–strain constitutive relationship. The reinforcement ratio primarily reflects the influence of the cross-sectional area of the reinforcement bars. From a practical perspective, efficiently improving the cracking resistance of a member under given working conditions is of greater concern. Therefore, this sensitivity analysis focuses on the influence of the concrete strength grade and the steel reinforcement ratio on the cracking moment. These two parameters are directly adjustable by the designer under given design conditions to enhance cracking resistance.

Again, the commonly used singly reinforced rectangular beam section in tests is selected: $b \times h = 150 \times 300$ mm, with the effective depth $h_0 = 270$ mm, and steel reinforcement of HRB400. Using the proposed method, the variation of the cracking moment with concrete strength grade is calculated for reinforcement ratios of 0.3%, 0.7%, 1.1%, and 1.4%. Similarly, the variation of the cracking moment with reinforcement ratio is calculated for concrete tensile strengths of 2.2 N/mm^2, 2.7 N/mm^2, 3.3 N/mm^2, and 3.9 N/mm^2. The results are shown in Figures 9 and 10. It can be observed that the cracking moment increases with either the concrete strength grade or the reinforcement ratio. A positive correlation exists between these two parameters and the cracking moment.

Figure 9 indicates that when the reinforcement ratio is below 1.1% (i.e., within the economical and low reinforcement ratio ranges), the cracking moment exhibits an approximately linear relationship with the concrete tensile strength. When the reinforcement ratio exceeds 1.1% (i.e., within the high reinforcement ratio range), the cracking moment initially increases at a lower rate with increasing concrete tensile strength within the range of lower concrete strength grades. As the concrete tensile strength increases into the range of higher concrete strength grades, the cracking moment gradually recovers to an increasing rate comparable to that observed under economical and low reinforcement ratios. Therefore, for beams with lower concrete strength grades and higher steel reinforcement ratios, the cracking moment is less sensitive to the concrete strength grade. Consequently, solely increasing the concrete strength grade is not the most efficient measure for enhancing the cracking resistance of the member under such conditions.

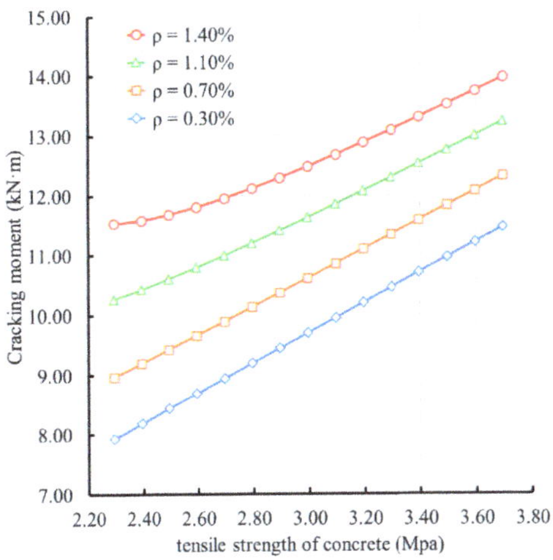

Figure 9. Variation of beam cracking moment with concrete tensile strength for different reinforcement ratios.

Figure 10 indicates that when the concrete tensile strength exceeds 2.7 N/mm^2 (i.e., within the higher concrete strength ranges), the cracking moment exhibits an approximately linear relationship with the steel reinforcement ratio. When the concrete tensile strength is below 2.7 N/mm^2 (i.e., within the lower concrete strength range), the cracking moment initially increases linearly with increasing reinforcement ratio within the low and economical reinforcement ratio ranges. As the reinforcement ratio increases into the high reinforcement ratio range, the cracking moment increases significantly faster. Therefore, for beams with higher reinforcement ratios and lower concrete strength grades, the cracking moment is highly sensitive to the reinforcement ratio. Consequently, increasing the steel reinforcement ratio is an effective measure for enhancing the cracking resistance of the member under such conditions.

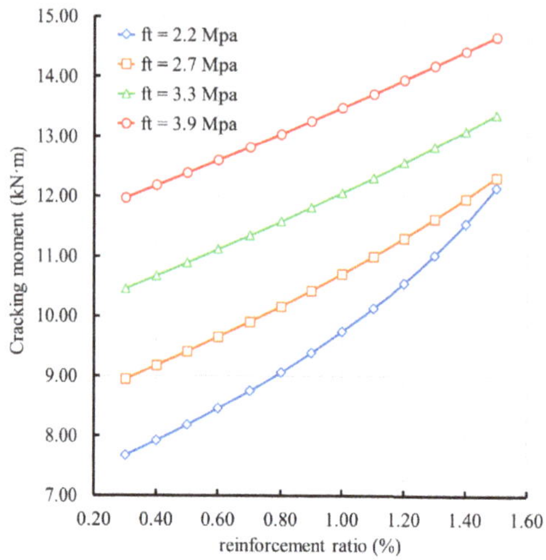

Figure 10. Variation of beam cracking moment with reinforcement ratio for different concrete strength grades.

In summary, within the range of economical reinforcement ratios (ρ = 0.7~1.1%) and normal- to high-strength concrete (f_t = 2.7~3.9 N/mm^2), the beam cracking moment always exhibits a linear relationship with both the concrete strength grade and the steel reinforcement ratio. The cracking resistance of the member can therefore be effectively enhanced by increasing either the concrete strength grade or the steel reinforcement ratio. However, when the concrete strength grade is relatively low and the steel reinforcement ratio is high, the contribution of the reinforcement to the sectional cracking resistance substantially exceeds that of the concrete. Therefore, under such conditions, increasing the steel reinforcement ratio is more effective for improving the cracking resistance of the member than increasing the concrete strength grade.

5. Conclusions

This paper proposes a strain energy-based method for calculating the cracking loads of steel- and FRP-reinforced concrete members, applicable to axially tensioned members and ordinary flexural members where tensile deformation is the primary deformation mode. The method establishes an expression for the total strain energy of the member using the axial tensile deformation as the independent variable. The critical condition for cracking is defined as the point where the second-order differential of this total strain energy with respect to the axial deformation equals zero. Based on this condition, the deformation at cracking and the corresponding cracking loads are determined. Compared to traditional calculation methods, the proposed approach requires only basic sectional geometrical parameters and the fundamental stress–strain constitutive relationship of the materials. It eliminates the need for introducing additional coefficients to describe the development of sectional inelastic deformation, such as the inelastic influence coefficient γ of the section resistance moment, making it simpler to apply. In contrast to emerging machine learning approaches, the proposed method possesses a clear physical basis and a transparent calculation process, ensuring the results are fully credible. In the derivation process, the constituent materials of the member are assumed to be homogeneous and isotropic. However, since the concrete stress–strain constitutive relationship adopted in this method is entirely based on experimental measurements, the results inherently incorporate the effects of material inhomogeneity and anisotropy. Therefore, the aforementioned assumption does not hinder the application of the method to materials like concrete, which are inhomogeneous and not strictly isotropic.

Calculations verified against experimental data from four sets of tests on steel- and FRP-reinforced concrete flexural members demonstrate that the proposed method is fully applicable to both normal- and high-strength concrete, as well as to both ordinary steel and FRP reinforcement. All results agree well with the experimentally measured values, with a relative error of only 1% and a coefficient of variation of 0.12 for the calculated-to-measured ratio. However, verification against experimental data for flexural members made with recycled aggregate concrete and ultra-high-strength concrete indicates that the proposed method is not applicable to these two types of concrete. For recycled aggregate concrete, the reduction in deformation resistance leads to a tensile stress–strain curve that does not match the curve based on conventional natural aggregate concrete used in this method. For ultra-high-strength concrete, the section remains in the elastic deformation stage at cracking, which contradicts the sectional inelastic deformation state assumed in the proposed method.

The proposed method is applied to analyze the cracking loads of axially tensioned and flexural members of steel-reinforced concrete. The results indicate that for axially tensioned members, when the reinforcement area exceeds a critical value (termed the inflection point reinforcement ratio ρ_D in this study), the member exhibits no significant loss of macroscopic performance, such as durability and stiffness, before the steel reinforcement yields or fractures in tension. Under this condition, the cracking resistance of the member increases severalfold. For flexural members, within the range of economical reinforcement ratios and normal- to high-strength concrete grades, the cracking resistance can be effectively enhanced by increasing either the concrete strength grade or the reinforcement ratio. However, when the concrete strength grade is relatively low and the reinforcement ratio is high, increasing the reinforcement ratio is more effective for improving the cracking resistance.

Owing to its foundation in traditional mechanics of materials, the proposed method can be readily integrated into existing structural design procedures. Furthermore, since it does not rely on any empirical coefficients that characterize the development of sectional inelastic deformation (such as the inelastic influence coefficient γ of the section resistance moment), there is no need to recalibrate such parameters through new tests for novel structural forms or materials. As a result, the proposed method not only avoids the complexity and uncertainty inherent in empirical parameters but also significantly enhances the adaptability of the cracking design approach to various new structures and materials.

Author Contributions: Conceptualization, methodology, software, validation, formal analysis, investigation, writing—original draft preparation, T.Z.; writing—review and editing, supervision, project administration, G.-Y.W. All authors have read and agreed to the published version of the manuscript.

Funding: This research received no external funding.

Data Availability Statement: Some or all data, models, or code that support the findings of this study are available from the authors upon reasonable request, including the experimental data and simulation results presented in this paper.

Conflicts of Interest: The authors declare no conflicts of interest.

References

1. Gao, D.; Zhao, J.; Zhu, H.; Zhang, Q. Calculation method for crack resistance of steel fiber concrete deep beams with reinforcement. *J. Hydraul. Eng.* **2002**, *33*, 124–128.
2. Li, Z.; Zhu, H.; Du, C. Experimental study on cracking behavior of steel fiber-reinforced concrete beams with BFRP bars under repeated loading. *Compos. Struct.* **2021**, *267*, 113878. [CrossRef]
3. Jia, B.; Zhu, S.; Liu, S. Theoretical Study on Cracking Moment of Normal Section of Concrete Beams Reinforced with FRP Bars. *Asian J. Chem.* **2014**, *26*, 5699–5702. [CrossRef]
4. Chen, X.; Zhang, Z.; Xu, Z. Experimental analysis of recycled aggregate concrete beams and correction formulas for the crack resistance calculation. *Adv. Mater. Sci. Eng.* **2022**, *2022*, 1466501. [CrossRef]
5. Zheng, W.; Lu, S.; Li, L. Experimental research on mechanical performance of reactive powder concrete beams reinforced with GFRP bars. *J. Build. Struct.* **2011**, *32*, 115–124.
6. Li, L. Mechanical Behavior and Design Method for Reactive Powder Concrete Beams. Ph.D. Thesis, Harbin Institute of Technology, Harbin, China, 2011.
7. Lu, S.; Zheng, W. Calculation method of normal section cracking resistance of GFRP reinforced reactive powder concrete beams. *J. Harbin Inst. Technol.* **2010**, *42*, 536–540.
8. Kong, X.; Han, S.; Gang, J.; Chen, X.; Wang, X.; Zhang, W. Study on flexural capacity of recycled concrete beams with GFRP bars. *Build. Struct.* **2024**, *54*, 50–56.
9. Saingam, P.; Chatveera, B.; Roopchalaem, J.; Hussain, Q.; Ejaz, A.; Khaliq, W.; Makul, N.; Chaimahawan, P.; Sua-iam, G. Influence of recycled electronic waste fiber on the mechanical and durability characteristics of eco-friendly self-consolidating mortar incorporating recycled glass aggregate. *Case Stud. Constr. Mater.* **2025**, *22*, e04369. [CrossRef]
10. Xu, S.; Zhao, G. A double-K fracture criterion for crack propagation in concrete structures. *China Civ. Eng. J.* **1992**, *25*, 32–38.
11. Xu, S.; Zhao, Y. Analysis and criterion of fracture process for crack propagation in concrete. *Eng. Mech.* **2008**, *25*, 20–33.
12. Zhao, Y.; Xu, S.; Wu, Z. A dual-G criterion for crack propagation in concrete structures. *China Civ. Eng. J.* **2004**, *37*, 13–18.
13. Xu, S.; Zhang, X. The new GR crack extension resistance as a fracture criterion for complete crack propagation in concrete structures. *China Civil Eng. J.* **2006**, *39*, 19–28+41.
14. Zhao, G. A new method for calculating crack resistance of hydraulic reinforced concrete with few bars and concrete structures. *J. Dalian Inst. Technol.* **1959**, *4*, 133–155.
15. Zhao, G. General calculation method for crack resistance of prestressed concrete, reinforced concrete and concrete members. *J. Civ. Eng.* **1964**, *2*, 1–16.
16. Cheng, W. Experimental study on crack resistance, stiffness and crack width of reinforced concrete rectangular section eccentric tensile member. *J. Nanjing Inst. Technol.* **1981**, *3*, 101–116.
17. Zhao, G. Calculation method for crack resistance (crack initiation and propagation) of hydraulic reinforced concrete structures. *J. Hydraul. Eng.* **1960**, *5*, 37–59.
18. *GB/T 50010-2010*; Code for Design of Concrete Structures. China Architecture & Building Press: Beijing, China, 2011.
19. *EN1992-1-1*; Eurocode 2: Design of Concrete Structures. British Cement Association: London, UK, 1994.
20. *ACI 318M-2014*; Building Code Requirements for Structural Concrete and Commentary. American Concrete Institute: Farmington Hills, MI, USA, 2014.
21. Guo, Z.; Zhang, X. Experimental study of the stress-deformation curve of tensile concrete. *J. Build. Struct.* **1988**, *4*, 45–53.
22. Guo, Z.; Shi, X. *Principles and Analysis of Reinforced Concrete*; Tsinghua University Press: Beijing, China, 2003.
23. Blackman, J.S.; Smith, G.M.; Young, L.E. Stress distribution affects ultimate tensile strength. *J. Proc.* **1958**, *55*, 679–684.
24. *CH10-57*; Chinese Translation of Soviet Code for Design of Prestressed Reinforced Concrete Structures. Architectural Engineering Press: Beijing, China, 1957.
25. Cai, S. Problem on the value of flexural and tensile cracking plasticity coefficient γs of concrete. *Concr. Reinf. Concr.* **1982**, *4*, 23–27+50.
26. Xu, J.; He, X. Preliminary analysis of volume (size) effect and anti-crack plastic coefficient γ value of concrete members. *Chin. J. Theor. Appl. Mech.* **1986**, *22*, 364–368.
27. Li, L.; Fan, X.; Shi, X. Experimental study on flexural behavior of large-scale prestressed UHPC-T-shaped beam. *China Civ. Eng. J.* **2018**, *51*, 84–94+102.
28. Huang, Q.; Hu, S.; Du, R. A calculation method for cracking moment of prestressed reactive powder concrete box girder. *China Civil Eng. J.* **2015**, *48*, 15–21.
29. Wang, Q.; Deng, J.; Gao, S. Study on the Applicable Calculation Method of the Cracking Moment of UHPC Beam. *J. China Three Gorges Univ. (Nat. Sci.)* **2023**, *45*, 70–76.
30. Lin, S.; Ji, B.; Xia, Z.; Yang, Y.; Lin, J.; Zhao, J. Crack resistance of prestressed RC-UHPC composite box girder. *J. Civ. Environ. Eng.* **2025**, *47*, 58–65.

31. Ali Talpur, S.; Thansirichaisree, P.; Poovarodom, N.; Mohamad, H.; Zhou, M.; Ejaz, A.; Hussain, Q.; Saingam, P. Machine learning approach to predict the strength of concrete confined with sustainable natural FRP composites. *Compos. Part C Open Access* **2024**, *14*, 100466. [CrossRef]
32. Ma, C.L.; Wang, W.H.; Hou, X.L.; Xie, C.X.; Lu, C.F. Prediction of the Shear Cracking Strength of RC Deep Beams Based on the Interpretable Machine Learning Approach. *J. Xinjiang Univ. (Nat. Sci. Ed.)* **2023**, *40*, 621–629.
33. Liang, Z.H.; Ying, Z.Q.; Liu, M.M.; Yang, S. Prediction Method for Reinforced Concrete Corrosion-induced Crack Based on Stacking Integrated Model Fusion. *J. Chin. Soc. Corros. Prot.* **2024**, *44*, 1601–1609.
34. Wang, Q.; Wang, T.; Xu, G. Comparative analysis of different calculation methods of cracking bending moment of reinforced concrete beams. *J. China Three Gorges Univ. (Nat. Sci.)* **2016**, *38*, 54–58+69.
35. Fang, S.; Zhang, P. An improved formula for cracking moment calculation of RC beams considering the development coefficient of plastic deformations. *Chin. J. Comput. Mech.* **2019**, *36*, 464–470.
36. Jiang, Y.; Liang, S.; Chen, D.; Lu, J.; Jiang, N. Experimental study on the deformation properties of high-strength reinforced high-strength concrete flexural components. *J. Build. Struct.* **1998**, *19*, 37–43.
37. Sun, M. Experimental Research on Cracking Moment of Reactive Powder Concrete Beams with High Strength Steel Bars. *Railw. Constr.* **2021**, *61*, 33–36.
38. *GB/T 50152-2012*; Standard for Test Methods for Concrete Structures. China Architecture & Building Press: Beijing, China, 2012.
39. Yuan, J.; Yu, Z. *Design Principle of Concrete Structure*; China Railway Press: Beijing, China, 2005.

MDPI AG
Grosspeteranlage 5
4052 Basel
Switzerland
Tel.: +41 61 683 77 34

Buildings Editorial Office
E-mail: buildings@mdpi.com
www.mdpi.com/journal/buildings

www.ingramcontent.com/pod-product-compliance
Lightning Source LLC
Chambersburg PA
CBHW061613210926
51850CB00009B/2311
9783725865567